The Greening of Sovereignty
in World Politics

The Greening of Sovereignty
in World Politics

edited by Karen T. Litfin

The MIT Press
Cambridge, Massachusetts
London, England

This book was set in Sabon by Doyle Graphics, Tullamore, Ireland.

Printed and bound in the United States of America on recycled paper.

Library of Congress Cataloging-in-Publication Data

The greening of sovereignty in world politics / edited by Karen T.
   Litfin.
      p.   cm. -- (Global environmental accord)
   Includes bibliographical references and index.
   ISBN 0-262-12211-1 (hc. : alk. paper)
   1. Environmental policy.   2. Sovereignty.   I. Litfin, Karen.
   II. Series: Global environmental accord.
   GE170.G737   1998   98-24300
   363.73′05--dc21   CIP

for Maya
and all the children

Contents

Series Foreword

A new recognition of profound interconnections between social and natural systems is challenging conventional constructs as well as the policy predispositions informed by them. Our current intellectual challenge is to develop the analytical and theoretical underpinnings crucial to our understanding of the relationships between the two systems. Our policy challenge is to identify and implement effective decision-making approaches to managing the global environment.

The Series on Global Environmental Accord adopts an integrated perspective on national, international, cross-border, and cross-jurisdictional problems, priorities, and purposes. It examines the sources and consequences of social transactions as these relate to environmental conditions and concerns. Our goal is to make a contribution to both the intellectual and the policy endeavors.

Preface

Observations about the incongruity between the political world, delineated by territorial boundaries, and the natural world, made up of interconnected ecosystems, have become something of a truism. The implication is that there is some fundamental incompatibility between sovereignty and ecology. Moreover, the truism takes on a tone of urgency as the scale and scope of environmental degradation expand: If our basic political institutions are by definition not up to the task, then what is to be done? If sovereignty and ecology are irreconcilable, then, in the name of survival, the first must be renounced for the sake of the second.

Yet the proliferation of international environmental agreements and transnational activism over the last three decades raises the possibility that existing political institutions, including the prevailing norms of sovereignty, can be altered in ways that permit and even foster ecologically benign practices. Sovereignty has proven itself to be an enduring and malleable set of norms, with its locus shifting from the absolute monarchs of the early modern period to "the people" in contemporary democracies. Thus, it is not surprising that we find the norms of sovereignty shifting once again in the face of attempts to cope with ecological destruction. The authors of this volume refer to this phenomenon as *the greening of sovereignty*.

To find evidence for and to theorize about the greening of sovereignty is not to suggest, however, that the world is ineluctably marching forward on a single path toward a paradise of sustainability. The greening of sovereignty is an uneven, variegated process, not a homogeneous one, and certainly the endpoint is not foreordained. Nor do arguments about the greening of sovereignty entail that all social actors are equally benefited by such a shift. The greening of sovereignty, as the authors in this volume demonstrate, is a highly contested social process.

The original impetus for this book dates back to a series of workshops convened by Ken Conca and Ronnie Lipschutz at the University of California at Berkeley, which led to the publication of *The State and Social Power in Global Environmental Politics* in 1993. My chapter in that book suggested that efforts to cope with global environmental degradation appeared to be provoking a kind of tug of war in which states' control and authority are paradoxically both bolstered and undercut. Conca expanded upon this argument in his brief and insightful 1994 essay in *Millennium*, as did Peter Haas and Jan Sundgren in their 1993 chapter in Nazli Choucri's *Global Accord*.

Despite these initial forays, there has been no concerted effort in the recent literature to examine and unpack the environment/sovereignty nexus. The gap is especially conspicuous given the burgeoning literature in the 1990s that seeks to reconceptualize sovereignty in the face of diverse challenges, ranging from finance and trade interdependencies to the rise of "ethnic conflict" in the aftermath of the Cold War. While much work remains to be done, this book begins to fill that gap.

The book grows out of a workshop held in Seattle at the University of Washington in 1995. That workshop, Rethinking Environment and Sovereignty, was made possible by a grant from the Social Sciences Research Council and the MacArthur Program in International Peace and Security. I am grateful to the Council and the MacArthur Program for their financial support, as well as to the Department of Political Science at the University of Washington for administrative and logistical support. I am also grateful to the workshop's five discussants and commentators who gave generously of their time and insight: James Caporaso, Joel Migdal, Janice Thomson, Rob Walker, and Mark Zacher. Their comments served as the basis for subsequent revisions of the papers.

I also wish to thank Clay Morgan at the MIT Press for his enthusiastic support for this project, as well as two anonymous reviewers who supplied detailed comments on an earlier draft of the manuscript. Of course, my deepest debt of gratitude goes to the authors. It has been a tremendous pleasure to work with such a diverse, intelligent, and innovative group of scholars, and I dearly appreciate the support they have given me throughout the project.

Karen T. Litfin
Seattle, WA
November 1997

Contributors

Daniel Deudney is assistant professor of political science at Johns Hopkins University. He has written extensively on environmental, energy, nuclear, space, and global governance issues and is coeditor of the forthcoming book *Contested Ground: Security and Conflict in the New Environmental Politics.*

Margaret Scully Granzeier is lecturer and program administrator in the Environmental Studies Program at the University of Chicago. She is presently involved in research projects on sovereignty and environmental protection and on environmental and cultural rights.

Joseph Henri Jupille is a doctoral candidate in political science at the University of Washington. His research interests include international environmental politics, institutions, and European integration.

Sheldon Kamieniecki is professor and chair of political science and director of the Environmental Studies Program at the University of Southern California. He is editor of *Environmental Politics in the International Arena: Movements, Parties, Organizations, and Policy* (1993) and coeditor of *Flashpoints in Environmental Policymaking: Controversies in Achieving Sustainability* (1997).

Thom Kuehls is associate professor of political science at Weber State University. He is the author of *Beyond Sovereign Territory: The Space of Ecopolitics* (1996).

Ronnie D. Lipschutz is associate professor of politics at the University of California at Santa Cruz. He is the author of *When Nations Clash: Raw Materials, Ideology, and Foreign Policy* (1989) and *Global Civil Society and Global Environmental Governance: The Politics of Nature from Place to Planet* (1996).

Karen T. Litfin is assistant professor of political science at the University of Washington. She is the author of *Ozone Discourses: Science and Politics in Global Environmental Cooperation* (1994) and recipient of a National Science Foundation grant to study the impact of changes in earth observation satellite technology on the international knowledge structure.

Marian A. L. Miller is associate professor of political science at the University of Akron. Her book *The Third World in Global Environmental Politics* (1995) won the International Studies Association's 1996 Harold and Margaret Sprout Award for best book on international environmental politics.

Ronald B. Mitchell is assistant professor of political science at the University of Oregon. His book *Intentional Oil Pollution at Sea: Environmental Policy and Treaty Compliance* (1994) received the International Studies Association's 1995 Sprout Award.

Paul Wapner is associate professor in the School of International Service at American University. His book *Environmental Activism and World Civic Politics* (1996) was awarded the International Studies Association's 1997 Sprout Award.

Veronica Ward is associate professor of political science at Utah State University. Her current research in international relations and social/public choice focuses on cooperative arrangements at both the domestic and global levels.

Franke Wilmer is associate professor of political science at Montana State University. She is the author of *The Indigenous Voice in World Politics* (1993), and her current research is on links between exclusionary identities and political mobilization for conflict in the recent Balkans war.

1

The Greening of Sovereignty:
An Introduction

Karen T. Litfin

Sovereignty and Environment: Beyond Either/Or

Recent years have witnessed the flowering of two parallel bodies of literature: one on global environmental politics and the other delving into the theoretical constitution of sovereignty. For the most part, the two have not been in communication with one another. Within the burgeoning field of international environmental politics, sovereignty, taken to denote the state's exclusive authority within its territorial boundaries, is often assumed to be "eroded" by efforts to address transboundary environmental problems. The geological metaphor is telling, for it conjures up a rather ahistorical and naturalistic image of sovereignty as a solid edifice. Among those with a more specific interest in conceptually unpacking sovereignty, environmental issues have been tangential concerns, but are typically presumed to pose a nebulous sort of challenge to sovereignty for similar reasons.

This book aspires to construct a bridge between these two literatures, suggesting that the relationship between environment and sovereignty is by no means a straightforward one. The authors have chosen to highlight the environment/sovereignty nexus because we believe that it embodies and elucidates some of the core issues and challenges facing world politics as we approach the twenty-first century. Moreover, we maintain that global ecology provides a vital site for uncovering the conceptual underpinnings and practices associated with sovereignty.[1]

While states may claim sovereignty over the resources and activities within their territories, they have come under mounting pressure to manage their resources according to internationally agreed upon norms. In the face of Chernobyl and looming global climate change, domestic energy policies are becoming increasingly answerable to international standards. In the face of an international outcry over tropical deforestation and new conditions attached to aid packages, developing countries have been compelled to alter their forestry practices. In the face of global ozone depletion, the international community has effectively outlawed a host of chemicals that were seen until recently as essential to industrial development. These examples, along with many others, evoke the drama and high stakes involved in the interplay of ecology and sovereignty, suggesting that traditional norms of sovereignty are being challenged by emerging environmental practices.

By virtue of its character as a multifaceted and historically variable concept and practice, sovereignty should be unbundled, and the intersection of environment and sovereignty provides a key site for that unbundling. We thus move away from the sterile question of whether efforts to cope with ecological problems erode or bolster some reified conception of sovereignty, to the more interesting question of how such efforts lead to the reconfiguration of political space. The state is unlikely to be placed on the endangered species list anytime soon, but sovereignty nonetheless seems to be undergoing a transformation in response to a host of functional problems and interdependencies, including ecological ones. It is the consensus of the authors in this volume that we are tracking a moving target; consequently, we aim to point out trends and tendencies rather than to make sweeping predictions or draw grand conclusions.

The composite and variable nature of sovereignty makes it a slippery concept, one which calls for disaggregation as a prerequisite to empirically grounded theory building. One might be tempted to eschew the term altogether, despite its ubiquity in international law and public discourse. But such a solution would only serve to drive a wedge between the worlds of theory and practice. A more fruitful strategy is to unpack the term, which has become a common practice among international relations theorists in recent years.[2] In unbundling sovereignty, we find various overlapping elements, each capable of further unbundling: territory, autonomy, authority, control, and population. Moreover, sovereignty entails rights, capacities, and responsibilities in three realms: those under a state's jurisdiction, those under other states' jurisdictions, and those under common jurisdiction. As the chapters in this volume suggest, each of these elements, in all of these realms, is being modified by political responses to transnational and global environmental processes.

The most widespread interpretation of the environment/sovereignty nexus is the popular "erosion-of-sovereignty" thesis. With "the seamless web of nature" standing in apparent contradistinction to the manmade patchwork system of nation-states, efforts to cope with international environmental degradation are frequently construed as a prima facie challenge to state sovereignty. Since sovereignty, the constitutive principle of the nation-state system, is premised upon territorial exclusivity, it is surmised that transboundary environmental problems necessarily undermine state sovereignty. Much of the early literature of the 1970s linked transnational environmental interdependence to the demise of the state system and anticipated its eventual replacement by some far-reaching supranationalism.[3] Alternatively, some predict the erosion of sovereignty from below, with power and authority devolving from the state to local communities whose identities and livelihoods are more entwined with specific ecosystems.[4] The proliferation of community-based environmental nongovernmental organizations (NGOs) is cited as evidence for this position.

The erosion-of-sovereignty thesis is appealing on logical as well as normative grounds: Not only do ecological holism and territorial exclusivity appear to be mutually exclusive, but the state has been an agent or accomplice in wrecking untold ecological havoc in the modern era. Unfortunately, the thesis is more generally assumed than elucidated, in great part because the concept of sovereignty has remained until recently "essentially uncontested"[5]—not only in mainstream international relations theory but in discussions of global ecological interdependence. This book is part of an emerging intellectual renaissance, or awakening to be more accurate, that is focusing attention on the conceptual underpinnings and metapractices associated with sovereignty.[6] It is the contention of the authors in this volume that the familiar concept of sovereignty, expressed in the modern territorial state, is being unsettled by a variety of forces, including environmental ones. But we do not conclude that sovereignty is being eroded in any wholesale or homogeneous fashion.

Despite the appeal of the erosion-of-sovereignty thesis, ecological integrity and state sovereignty do not necessarily stand in opposition to one another. Some would argue, as I have suggested elsewhere, that only the state possesses sufficient authority, legitimacy, resources, and territorial control to enforce environmental rules and norms.[7] The negotiation, implementation, and enforcement of environmental treaties are largely in the hands of states. A decline of sovereignty might even undermine the ability of the state to comply with international obligations and actually protect the environment. Indeed, since state sovereignty can function as a bulwark against opportunistic economic forces, some environmental regimes have sought to enhance state capacity, particularly in developing countries.[8] Moreover, challenges to state sovereignty may be environmentally harmful, as in the 1994 GATT ruling that the United States had no sovereign right to boycott tuna caught with fishing methods that killed large numbers of dolphins.[9] As we see, sovereignty need not be an adversary of ecology. Even among those who might claim that state sovereignty promotes ecological sustainability, however, the core notion of sovereignty remains untheorized.

While each position has its merits, together they lead to incompatible perspectives on the relationship between environment and sovereignty. This is primarily because both lines of reasoning rely upon a monolithic understanding of sovereignty as something that is either eroded or fortified. As Ken Conca suggests, both positions "fail to disaggregate what is in fact a complex and highly unevenly distributed set of international pressures on states to solve environmental problems."[10] The scope of state autonomy may be narrowed by pressures from above and below, as the erosion-of-sovereignty thesis claims, even as the problem-solving capacity of states increases, as the sovereignty-as-bulwark perspective submits. Or a state's control, autonomy, and authority can be enhanced at home, even as its autonomy to act in the global commons is constrained. The relationship between environment and sovereignty is not a straightforward one.

Once sovereignty is understood as a socially constructed institution that varies across space and time, with multiple meanings and practices that are not set in stone, then its relationship to environmental questions can be more easily studied. Social construction means that social agents produce, reproduce, and redefine the constitutive principles and structures by which and in which they operate.[11] For our purposes, an emphasis on social construction need not exclude material forces; indeed, for the authors in the volume, there seems to be a dialectical interplay between environmental forces and the social norms associated with sovereignty. Despite the naturalistic imagery commonly invoked in discussions of sovereignty, few would counter the claim that sovereignty is a socially constructed practice. Yet the real meaning of that claim only emerges when we consider the elements of sovereignty, not with an eye to erecting a fixed taxonomy or a new master definition, but in order to understand better the various claims that arise from its multifaceted character. In unbundling sovereignty, we find a host of overlapping elements, each of which is socially constructed and requires further unbundling, including territory, autonomy, nonintervention, recognition, authority, and popular sovereignty. In this section, I offer a brief overview of how sovereignty is understood in the field in order to draw out the major traits associated with it.[12] This process of unbundling serves as the basis for the conceptualization of sovereignty adopted in this volume in terms of autonomy, control, and authority.

Unbundling Sovereignty

The classic definition of sovereignty given by F.H. Hinsley as "final and absolute authority in a political community"[13] fosters a monolithic conceptualization of sovereignty. Yet even here, as both principle and practice, sovereignty is commonly understood as having two dimensions: external and internal. As a principle, *external sovereignty* refers most narrowly to the state's legal or constitutional independence vis-à-vis other states.[14] This conceptualization informs the United Nations's principle of "sovereign equality," which, like property, is a seemingly absolute concept: An entity either has it or does not have it.[15] On this reading, sovereignty functions as the gate between domestic politics and international relations. Domestically, it refers to the absolute authority within the political community; internationally, it entails the antithesis of anarchy, where no supreme authority exists. Hence, only states can formulate foreign policy and engage in diplomacy; sovereignty here is seen as the objective characteristic that entitles states to engage in international relations. Therefore, legal sovereignty, a key element of external sovereignty, is paradoxically reinforced by international law, including environmental treaties, even as states' autonomy of action is circumscribed, because only states can be parties to treaties. In legal terms, then, every international environmental agreement simply fortifies and reproduces the constitutive principle of sovereignty.

Despite its apparently fixed and objective status, however, legal sovereignty is socially constructed, for it is a form of legitimation through which power is converted into authority. As Hinsley himself recognized, *territorialization*, the key to the integration of political power and authority with community, was vital to the emergence of modern notion of sovereignty.[16] The territorial exclusivity of the modern state was a social construction which took centuries to achieve. Legal sovereignty is constructed through a process of *recognition*, whereby putative states must engage in certain practices and establish acceptable institutional structures in order to be admitted to the club of sovereign states. The salience of this process of recognition was recently evident in the breakup of the former Soviet empire. The standards of recognition and legitimation, along with the social meaning of territory, are malleable under changing conditions.

The meaning of *authority* is also contestable. International relations theorists, starting with a concern with interstate rather than state-society relations, typically adopt a state-centric and militarily oriented perspective, emphasizing "the state's dependence on other states for its authority" and the state's recognized right to "a monopoly on coercive activity in its territory."[17] Other social theorists, drawing on the Weberian tradition, would link authority to social forms of legitimation.[18] As the number and significance of transnational actors and linkages, as well as subnational political identities, have grown in recent years, international relations specialists have been more willing to incorporate social conceptions of legitimacy into their understandings of authority and sovereingty.[19] Global ecological interdependence, thrust onto the world stage by a host of nonstate actors, is a key development in the transformation of patterns of international authority.

Internal sovereignty, a logical corollary of legal sovereignty, is typically understood as the state's autonomy over its own affairs. Here *autonomy* may refer either to the state's recognized *right* to order its own affairs, or its actual ability to *control* people and processes within its territorial boundaries. Essential to internal sovereignty, therefore, is the principle of *nonintervention*: State X's sovereignty is guaranteed by the norm against intervention by foreign states in X's internal affairs. Sovereignty, then, is seen to function as a kind of dividing line between inside and outside, domestic and international, with territorial boundaries serving as the physical expression of that dividing line in nature.

A state may enjoy legal sovereignty, however, while having very little *operational (or effective) sovereignty*. Rights and principles do not necessarily entail capabilities. The gap between sovereignty as principle and practice is nowhere greater than in the Third World. Robert Jackson's provocative notion of "quasi-states" rests upon the distinction between "negative" sovereignty as constitutional independence, and "positive" sovereignty as the capacity to act.[20] This distinction has far-reaching implications for environmental practices. A quasi-state may agree to a treaty or establish an environmental agency, for instance, but that says nothing about its ability to monitor and enforce environmental standards. Many "sovereign" Third World states even lack sufficient funds to send delegates to international treaty negotiations. Thus, a formalistic understanding of sovereignty as constitutional independence offers little insight into the environment/sovereignty nexus, either in the Third World or elsewhere. A pragmatic conceptualization that focuses on the positive capacity to act, or *operational sovereignty*, is more illuminating.

Especially for realists, *sovereignty* is understood as the bedrock principle underlying the Westphalian system of mutual rights and responsibilities, a system which realists take as resilient and enduring, but which others argue is under attack from a host of global interdependencies.[21] While the ideal typical world of legally and politically autonomous units has never existed in practice, the chapters in this volume suggest that it is increasingly questionable in a world in which states' autonomy, control, and authority are limited not only by other states and international institutions, but by a growing array of nonstate economic, social, and political actors. Most of the chapters do not argue, however, that sovereignty is on the verge of extinction, but rather that international environmental responses are "greening" sovereignty, in some arenas more than others and in some places with more positive consequences than others.

Since constitutive principles are only modified through process, and since it is difficult to think of any instances where legal sovereignty has been compromised by ecological issues, the chapters in this volume focus on operational sovereignty, conceptualizing sovereignty as a changing practice involving patterns of autonomy, control, and authority. Historically, it is an institution with European origins whose locus has devolved from God (and his delegate in Rome) in the Middle Ages, to the monarchs of the absolutist period, to the bureaucratic, territorial, and (more or less) democratic state associated with modernity.[22] In its modern incarnation, sovereignty is essentially about dividing inside from outside, and exercising authority, autonomy, and control over that which transpires within. This formulation suggests that there are no fully sovereign states, if indeed there ever were, and that environmental concerns are likely to present further challenges to the practices associated with sovereignty. If sovereignty is about divisibility, then ecological indivisibility—or even the perception of it—necessarily calls into question traditional notions of independent decision making, territorial control, and exclusive authority. But a challenge to existing practices does not require their disappearance; sovereignty has proven itself to be a durable institution, capable of mutating to survive in diverse cultural and economic settings over the centuries. Just as the absolutist sovereignty of seventeenth-century Europe differed greatly from later liberal notions and practices, the social and economic transformations of the late twentieth century, along with their ecological reverberations, appear to be reconfiguring the meaning of sovereignty.

The Tripartite Division of Sovereignty

Although the precedent for disaggregating sovereignty is established, the tripartite division proposed here—autonomy, control, and authority—requires some justification. As we have seen, various schools of thought tend to emphasize one or the other of these dimensions of sovereignty, a tendency that explains at least part of their theoretical dissent. Those who define sovereignty in terms of legal authority, or constitutional independence, are likely to conclude that environmental challenges result in no significant reconfiguration of sovereignty. Those who conceptualize authority more broadly in terms of legitimacy and patterns of governance, like those who emphasize the dimensions of autonomy and control, are more likely to find environmental practices leading to new norms of sovereignty.

This tripartite conceptualization of sovereignty has several advantages. First, it places the emphasis clearly on operational sovereignty, so that actual practice rather than legal formalism becomes the focus of empirical research. Given that constitutive principles are only modified through changing practices, this orientation makes sense. Yet this formulation does not exclude legal sovereignty; it simply includes it as a subset of authority. Nor does this reading of authority privilege the state's right to make an authoritative decision to go to war. Such a reading would offer little insight into the impact of environmental practices on changing patterns of authority.

Second, this tripartite division is broad enough to encompass both internal and external sovereignty, without a priori privileging either. In fact, each of the three dimensions may include both an internal and an external dimension.

Third, conceiving of sovereignty in terms of autonomy, control, and authority usefully decenters the state. The history of sovereignty shows that it can be an attribute of various political entities, not just the state. Similarly, territoriality is essential to the modern notion of state sovereignty, but that does not make it integral to sovereignty in general. Given that entities other than the state have been and may be considered sovereign in the future, territory need not be associated with sovereignty. Moreover, the empirical issues associated with territoriality are encompassed by the dimensions of autonomy and control.

If we understand sovereignty as a historically variable amalgamation of autonomy, control, and legitimacy, then much of the literature on global environmental politics indicates that the referents of those terms are undergoing significant change.

The Environment/Sovereignty Nexus

The establishment of international environmental institutions and the activities of transnational environmental actors, particularly NGOs and scientists, are creating new forms of governance and authority. While these may not stand poised to replace the state, they may be modifying the character of sovereignty. Certainly neither the Westphalian system of rights and responsibilities nor its liberal successors was particularly sensitive to the notion of ecological responsibilities. Nature, as source and sink, was assumed to be bounteous—almost infinitely so. Today's emergent discourses of sustainability betoken a growing awareness of limits, compelling the practices of sovereignty to incorporate some sense of ecological accountability. This is consistent with Janice Thomson's important insight that sovereignty is not a static organizing principle, but rather a historical institution whose norms and practices have changed significantly over time. Just as states accepted responsibility for the violent actions of their citizens in the nineteenth century, leading to the virtual elimination of piracy and mercenarism, so too are states today beginning to take responsibility for ecologically harmful activities that emanate from their territories.[23] If environmental action leads to meaningful shifts in patterns of authority, then we should expect the practices associated with sovereignty to be revised. The perceived significance of this reconfiguration for the character of political life, however, varies greatly across schools of thought.

According to liberal institutionalists, states cooperate to cope with environmental problems by creating new international regimes and organizations.[24] These new institutions may decrease states' autonomy of action, but they always reinforce their legal sovereignty and very often enhance their problem-solving capacity as well. Modified practices of sovereignty do not entail any fundamental political transformation. As Peter Haas and Jan Sundgren conclude in their study of the impact of international environmental law on the practices of sovereignty, "These new practices of sovereignty have emerged without a tranformation of the juridical principle of sovereignty or of the international anarchic order within which countries exercise choice. Revolutionary change occurs through reformist means."[25] The alleged revolution is apparently the greening of state practices, not any transformation of state sovereignty.

For institutionalists, environmental concerns may be generating forms of political activity beyond the state, but states remain the key players within the new international institutions. On this view, states engage in *sovereignty bargains*, in which some limitations on their autonomy which ultimately enhance the effectiveness of sovereignty are voluntarily accepted.[26] Yet other observers note that politics beyond the state can be a primary factor in forging these sovereignty bargains. With growing citizen activism in response to the expanding spatial and temporal scope of environmental degradation, sovereignty bargains in the form of international environmental regimes have proliferated. The cumulative effect of such bargains, however, may be to alter the norms and practices of sovereignty, and perhaps create and legitimize alternative channels of political activity.

A potentially more far-reaching interpretation of the reconfiguration of sovereignty is found in the writings of Ronnie Lipschutz, Paul Wapner, and others on social movements, NGOs, and "global civil society."[27] Here, the emphasis is not so much on international environmental treaties and institutions, but on "politics beyond the state," or the formation of alternative channels of control and authority and the reshaping of social meanings and beliefs by nonstate actors. As Hugh Dyer argues, although these shifts are occurring at a cultural level, they may have profound political implications.[28] Yet it is important to recognize that these shifts in control and authority, along with changing social meanings, can occur as a result of international regime dynamics as well. While states may consciously make certain tradeoffs among authority, control, and authority when they engage in sovereignty bargains, those bargains may also engender unintended consequences.

International environmental regimes, for instance, alter the function of boundaries by "untying the bundle of rights conventionally associated with full territorial sovereignty."[29] These rights are fundamentally about autonomy, control, and authority within a territory. The meaning of *territory*, along with its place in the set of practices associated with sovereignty, is being modified by environmental responses. If territory provides the container for state sovereignty, then transnational environmental problems and efforts to address them seem to be reshaping that container. John Ruggie suggests that the practices of international custodianship implicit in efforts to address global ecological interdependence may be producing an embryonic "multiperspectival" polity which, while not presently threatening to replace the existing system of rule, nonetheless stands in sharp contrast to the single-point perspective entailed in territorial exclusivity.[30] Many of the essays in this volume support Ruggie's hypothesis, but some do not.

A close reading of environmental practices highlights the fact that territory itself is socially constructed, and hence cannot be unbundled; thus territory should not be assumed as a taken-for-granted container for state sovereignty. As Thom Kuehls argues in chapter 2, territories are constructed through a series of disciplinary maneuvers that make them available as resources to the state. The social construction of *territory* among indigenous peoples, for instance, is strikingly different from its construction by the state. Territory in sovereign states is equated with natural resources available for economic or geopolitical exploitation. The instrumental dimension, however, is only part of the meaning of territory for indigenous peoples, and it is typically overlaid with a complex web of cultural and religious practices which underscore their enduring understanding of it as sacred.[31] Territory, therefore, should not be understood simply as an empty container for state sovereignty, but as intersubjectively constructed and potentially negotiable. The assignment of ecological responsibility for harmful processes that emanate from within the borders of "sovereign" states, articulated in recent international regimes, confirms the assertion that territorial norms can be renegotiated.

Likewise, the principle of *nonintervention*, which is essentially about a state's autonomy delineated by territorial boundaries, is being modified in the face of environmental considerations. Traditionally, *intervention* has been understood in terms of military force, but countries can also be "invaded" by foreign pollutants or even by the depletion of a crucial resource. The technological, economic, and ecological exigencies of the late modern period, along with the material weaknesses of many Third World states, have generated new forms of intervention that elude conventional categories—including a host of transboundary pollutants and interventionist environmental policies. Thus, the meaning of nonintervention, like that of territory, is being renegotiated in the context of global ecological interdependence.

Some transboundary pollution problems have been framed in terms of foreign intervention, and can be viewed as clashes among competing sovereignty claims. Sweden, for instance, claimed that the death of its lakes in the 1960s and 1970s constituted a form of intervention by Britain and West Germany, whose export of sulfur dioxide emissions was linked to acid precipitation. Meanwhile, Britain defended its "sovereignty" until the mid–1980s, insisting that any international efforts to regulate sulfur emissions represented a form of foreign intervention. Once the issue was resolved in favor of the Nordic countries, largely because of intense pressure from transnational environmental groups, the meaning of intervention had been expanded to include the export of air pollution.

The most obvious example of intervention in environmental affairs by a supranational organization is the European Union (EU), which has moved in recent years to harmonize environmental standards among its members. As Joseph Jupille argues in this volume, the EU's intervention in domestic environmental matters, and thus its infringement on state sovereignty, has not been as dramatic as one might imagine. Yet, although states continue to enjoy extensive autonomy and control in the implementation of EU directives, the emerging principle of subsidiarity entails an alternative to traditional sovereignty as an approach to the delineation of political authority.

Concerns about foreign intervention are heard most frequently from developing countries, many of which suspect that the newfound ecological concern of industrialized countries is merely the latest chapter in a long history of imperialism. This conviction is reinforced when environmentalists assert that China should not develop its coal resources, that Kenya must protect the rhino, or that tropical countries must stop destroying "our" rainforests.[32] The principle of nonintervention is increasingly applied to nonstate actors as well, including foreign media and NGOs. The Brazilian government, for instance, complained in the late 1980s that foreign television broadcasts on tropical deforestation were part of "an orchestrated campaign for the internationalization of the Amazon."[33] Similarly, governments in developing countries often claim that the activities of foreign-based NGOs violate their sovereignty, particularly when their environmental work is tied to human rights advocacy. Their concerns are simultaneously about autonomy, control, and authority.

The multilateral development banks (MDBs), which to a great extent set the tone for Third World development, may be the most influential supranational agencies with respect to environmental issues. For years, developing countries have complained that the conditions attached to MDB loans constitute a violation of their sovereignty.[34] This argument has been revived as new forms of "green conditionality," often promoted by northern NGOs, have been attached to MDB loans. Similarly, proposed debt-for-nature swaps have been criticized as eco-imperialist, just as the compulsion of debt-ridden countries to plunder their natural resources in order to earn hard currency may be construed as an indirect form of foreign intervention. In each case, the meaning of nonintervention is highly contentious and open to revision. Developing countries have declared that sovereignty includes "the right of states to use their own natural resources in keeping with their own developmental and environmental objectives and priorities."[35] This is simultaneously a statement about the right to nonintervention by foreign powers (autonomy) and the right to domestic control. But it is about rights, not capabilities; even if states have the will to work toward ecological sustainability, they may lack the ability to, say, internalize environmental costs by altering the tax structure or intervening otherwise in markets.

Like territory and nonintervention, *internal sovereignty* may be unbundled and its practical meaning altered in the face of ecological challenges. Ken Conca distinguishes two dimensions of internal sovereignty, *capacity* and *legitimacy*, which in many cases are inversely related. These two dimensions, it should be noted, are subsets of control and authority, respectively. Using the case of Brazilian sovereignty over the Amazon, Conca finds that even as external sovereignty is narrowed in scope by international prohibitions, state capacity may be deepened as the state increases its ability to penetrate civil society, while state legitimacy is concurrently eroded.[36] Those states with the capacity to exert domestic control may use environmental conservation as a subterfuge to promote preexisting statist agendas, as in Kenya's wildlife protection policies and Indonesia's state forestry policies.[37] When states destroy the natural environments of native peoples, like the "national sacrifice area" in the Four Corners region of the United States and the mammoth hydropower projects along India's Ganges and Narmada rivers, such exploitation becomes a form of "intrastate colonialism."[38] But since genuine authority requires legitimacy, and not merely physical control, a coercive state may ultimately undercut its own authority. Very often, the victims of "coercive conservation" are indigenous peoples.

When states become environmental predators, or when they lack the will or the ability to solve environmental problems, nongovernmental actors may seek to fill the void, highlighting another dimension of sovereignty linked to authority: *popular sovereignty.* The degree of tension between state sovereignty and popular sovereignty is inversely related to the level of participatory democracy. Where governments are hostile to NGOs, foreign funding for NGOs in developing countries has become a thorny political issue, with governments framing the issue in terms of sovereignty and sometimes even prohibiting or restricting the flow of foreign funds to local NGOs. Those funds can be quite substantial; in 1989, NGOs from the North distributed about $6.4 billion to developing countries—about 12 percent of all public and private development aid.[39] If official assistance continues to decline, the relative importance of private aid will increase, further complicating the relationship between state and popular sovereignty in developing countries.

Because democracy traditionally has been about processes within the state, international relations theorists have tended to ignore the popular dimension of sovereignty. The task of theorizing about democracy has largely fallen to scholars of American politics and political theorists, while students of international relations have hung their hats on a simplistic and monolithic image of sovereignty that more closely resembles Hobbes's despotic Leviathan than any contemporary democracy. The question of democratic participation is generally reduced to one of self-determination; once the instrumentality of the state is established, that is the end of the question. This tack ignores the fact that *the authority of the modern state rests not merely on coercion, but on legitimacy.*[40] More importantly, it reifies the state as the only viable form of community, relegating popular movements to the status of "prepolitical, merely social creatures."[41] The dramatic increase in transnational citizen activism suggests that the time has come for international relations scholars to consider the implications of democratic participation for sovereignty, and not just with respect to environmental issues.

Although international treaties and intergovernmental organizations have become increasingly important in addressing environmental problems, the driving force behind virtually all of these apparently state-centric activities has been popular pressure expressed through NGOs.[42] Rarely does the state, even in its most environmentally sensitive agencies, take a proactive ecological stance—either internationally or domestically. NGOs are the driving force behind efforts to cope with ecological problems, at every level of social organization. Moreover, the intersection of environmental NGOs with development organizations, indigenous peoples, and human rights groups is indicative of the complex web being spun by global ecopolitics.

Evoking a broader, nonstatist notion of politics as employing "the means to order, direct, and manage human behavior" in the public realm, the literature on global civil society and world civic politics suggests that global ecopolitics is generating significant new forms of governance, primarily through certain cultural shifts associated with the diffusion of ecological sensibility. Thus, transnational environmental activities may be understood as attempts to revise the norms and practices of sovereignty. Global environmental politics deals with assigning collective responsibility for the health of the planet, not just at the level of the state but at all levels of social organization; claims about responsibility generally have implications for the political distribution of autonomy, control, and authority.

Finally, although sovereignty is generally understood in fairly narrow political terms as a defining feature of the state and the state system, it can also be interpreted more broadly as a cultural practice associated with modernity. Like any other cultural practice, then, it should not be torn assunder from the other forms of life with which it is associated. On this reading, sovereignty should be contextualized with other features of modernity: individualism, secularism, liberal democracy, science, specific readings of masculinity, and certain conceptions of linear time and striated space.[43] The unbundling of sovereignty as a political practice, then, would have implications for the larger cultural complex associated with modernity. Of all issues, ecological ones, with their grass roots constituency, their scientific underpinnings, their essential borderlessness, and their intergenerational temporal component, may be especially likely to alter the broader matrix of modernity's cultural practices. Several of the chapters in this volume suggest that the reconfiguring of political spaces, identities, and communities in the face of environmental concerns may be the only source of realignment within a larger cultural shift.

Overview of the Chapters

The contributors to this volume feel that it would be counterproductive, and probably impossible, to submit a singular definition of sovereignty. We believe that it is possible to carry on a coherent conversation about sovereignty without erecting a definitional hegemony. Despite the divergence on a formal definition, all of the essayists understand sovereignty to be multifaceted rather than monolithic, and thus capable of being unbundled. Moreover, each is aware that the project of rethinking sovereignty and environment requires tracking a moving target—a requirement that is a source of difficulty as well as theoretical inventiveness. Some of the authors approach sovereignty as a political principle and practice associated with the state; others understand it more broadly as a cultural practice characteristic of Western modernity. These diverse conceptual vantage points are not mutually exclusive. Indeed, each is capable of illuminating certain aspects of the sovereignty problematique that remain inaccessible to others.

Nor do we all share the same interpretation of *environment*. For some of us, "the seamless web of global ecological interdependence" is a natural phenomenon, confronting and even opposing the artificial boundaries erected by sovereign states. For many of the authors in this volume, however, environmental problems are themselves socially constructed, even when they have concrete—at times, even dramatic— physical ramifications. The social construction of nature, on this reading, is intimately connected with the social construction of sovereignty. In other words, the "sovereignty of nature" (see Lipschutz, Chapter 4) is socially constructed in ways that rationalize and mask relations of power and domination. Thus, this volume seeks to contribute not only to the theoretical exploration of *sovereignty*, but also to some extent *environment*.

We are an epistemologically diverse group of political scientists, and we believe that our diversity helps us to contribute meaningfully to an understanding of the environment/sovereignty nexus. Some of us are international relations scholars; others are political theorists. Some of us are institutionalists, some are influenced by social constructivism, poststructuralism, and normative theory. Consequently, we explore the impact of different dynamics on the environment/sovereignty nexus.

Mitchell and Miller frame their arguments in terms of issue framing; Jupille and Wapner in terms of international law; Wilmer, Kuehls, and Scully Granzeier and Kamieniecki in terms of culture; and Litfin in terms of technology. Yet we are united in our belief that the greening of sovereignty represents an important new dynamic in world politics, with potentially far-reaching theoretical implications.

The next four chapters look at some of the theoretical tensions between environment and sovereignty. Environmental problems are typically viewed as a challenge to sovereignty because of the penetrability of territorial boundaries, a view which takes territory and environment as unproblematically given. Thom Kuehls argues otherwise in Chapter 2. His exploration of governmentality, a concept developed by Foucault from Rousseau's work, shows that territories and populations are socially constructed as appropriate objects of sovereignty. This historicized notion of the sovereign state runs contrary to the abstract state hypothesized by neorealists. As an illustration of how populations and territory are constructed through governmentality, Kuehls analyzes an eighteenth-century document from Portuguese America. Relating his argument to more contemporary concerns, he argues that proposals by the World Commission on Environment and Development to address global environmental crises entail the same sorts of governmental practices that constructed the sovereign state, suggesting the need for alternative practices of governmentality.

While Kuehls emphasizes the construction of territory and population under the modern state, Franke Wilmer takes up the implicit claim that the social construction of nature among indigenous cultures is fundamentally different from that of "Western" cultures. In Chapter 3, Wilmer demonstrates that indigenous conceptions of nature, reason, and power generate radically different understandings of both environment and sovereignty among these cultures. The roots of these differences, she argues, are to be found in the construction of the self through self/other relations. Feminist psychoanalytic theory may provide some insight into the ways in which self/other and human/nature relationships are constructed among indigenous peoples and in Western cultures based upon their different familial structures. Wilmer argues that the obsession with separation and autonomy, characteristic of the development of the Western masculine self, has important implications for the practices associated with both sovereignty and environment. The inability of prevailing conceptions of sovereignty to satisfactorily address environmental exploitation, she concludes, is indicative of a serious underlying cognitive failure—the failure of the Western self to reach maturity.

Veronica Ward is interested in the conditions under which scientific discourse can contribute to the redefinition of sovereignty in the global commons. But she finds that the apparent contradiction between the borderlessness of nature and the world of sovereign states cannot always be resolved by science-based arguments. In Chapter 4, she explores the holistic, intergenerational principle of ecosystem management, which has emerged from conservation biology, in order to ascertain some of the conditions under which the apparent clash between sovereignty and ecological principles can be reconciled. Ward cites the Convention on the Conservation of Antarctic Marine Living Resources, the only international agreement in which ecosystemic management principles have been implemented relatively successfully. She finds that the relative ease of identifying ecosystem boundaries, along with the tangled web of sovereignty claims and denials of claims embedded in the Antarctic Treaty System, were instrumental to the application of ecosystem management there. These conditions, she concludes, are not likely to be duplicated elsewhere in the near future.

The social construction of nature and sovereignty proceed not just through the scientific and psychological discourses that Ward and Wilmer cite, but also through social and political discourses. Ronnie Lipschutz, in Chapter 5, argues that certain environmental discourses, in taking global ecological interdependence as a purely natural phenomenon, serve to reify certain relations of power. Attempts to draw on the "determinism" of nature, whether via science or ethics, are nothing new; the great geopoliticians of the late nineteenth and early twentieth centuries all sought to ground the practices of sovereignty in the relative immutability of nature. Contemporary discussions of "limits" and "sustainability," argues Lipschutz, are the direct descendants of geopolitics and, as such, assume boundaries where, perhaps, none should exist. In tracing the linkages between these two seemingly unrelated discourses of geopolitics and ecological interdependence, Lipschutz maintains that, as with geopolitics, the sovereignty of nature over human affairs is not as clear as it might seem. Sovereignty and ecological interdependence are both socially constructed, but by showing how the discourses in which they are embedded tend to naturalize them, Lipschutz is able to denaturalize them.

The four chapters in Part II examine the aspects of transformation and continuity of sovereignty in specific case settings. If the principle of territorial sovereignty has any enduring meaning at all, we should not be surprised that perhaps the most far-reaching reconfiguration in international patterns of control, autonomy, and authority has occurred in the global commons, as Ward's exploration of ecosystem management in Antarctica confirms. In Chapter 6, Ronald Mitchell examines the reconfiguration of sovereignty with respect to another global commons issue: whaling. He finds that international efforts to redefine sovereignty can been supported by three types of discursive constructions of nature: interest-based, scientific, and normative. Using the whaling case, he finds that science-based arguments led states to accept and abide by collective decisions that would constrain behavior previously considered a sovereign right more readily than interest-based or normative arguments. Mitchell's findings in the whaling case support the argument developed in the chapters in Part I: the ways in which "environment" is constructed have a real impact on the norms and practices of sovereignty.

By their very nature, environmental regimes represent a modification in the way states exercise sovereignty, a modification that Marian Miller argues in Chapter 7 is likely to be more significant for Third World countries than industrialized countries. She argues, however, that the specific property and access circumstances of the environmental issue can serve as mitigating factors. Miller's examination of the regimes for stratospheric ozone, hazardous waste trade, and biodiversity suggests that Third World countries are better able to salvage their effective sovereignty when the issue is a common property, open-access one such as ozone depletion. In such a case, Third World countries can exercise some modest influence in the formation of environmental regimes. But, in the absence of the perceived interdependence that is the consequence of common property circumstances, Miller argues that the industrialized world is likely to continue the transfer of economic and ecological costs to the developing world, thereby accelerating the drain of sovereignty from Third World states. Again, the discursive framing of an environmental issue, whether it is perceived as a commons or a transboundary problem, has a real impact on how sovereignty is reconfigured by efforts to address it.

This line of argument emphasizes that environmental problems are not just physical phenomena. In Chapter 8, I argue that they are also informational phenomena, and that the social construction of nature can be influenced by the technological instruments through which information is produced and disseminated. The fact that environmental information is increasingly derived from satellite-based instruments is reconfiguring sovereignty in some surprising ways. Like other technologies, earth-observing satellites may be explored as "artifact/ideas," which entail a logic of power and authority. But the logic of earth remote sensing is not always a clear and unidirectional one with respect to the norms and practices associated with sovereignty. While states are the primary originators and proprietors of satellite-based information systems, the resulting transparency and multiple channels of communication and information diffusion seem to be facilitating a shift away from the state's territorial exclusivity and informational preeminence.

Sovereignty is also affected by the institutional venue in which environmental policies are debated and decided. In Chapter 9, Joseph Jupille finds that the environmental agenda and institutional changes in the European Union are leading at least to discursive innovations, and perhaps innovations on the ground as well, with respect to three components of sovereignty: legal authority, power, and autonomy. His analysis reveals that EU member states' legal authority is on the wane as a result of environmental exigencies, but that their power and autonomy remain largely intact. Jupille argues further that the EU's principle of subsidiarity, according to which policy-making authority should reside at the level most appropriate to the problem being addressed, represents a novel approach to environmental protection which lies outside the sovereignty framework. While both the diminution of legal authority and the rise of subsidiarity are currently unique to the EU, Jupille suggests that these innovations may nonetheless be applicable to the intersection of sovereignty and environment more generally.

Jupille's thoughts on subsidiarity as an alternative principle of political order move us into Part III, "Revisioning Sovereignty." The chapters in this section present sequentially more encompassing renderings of how sovereignty is being, and might in the future be, reconfigured by environmental practices. In their chapter, Sheldon Kamieniecki and Margaret Scully Granzeier argue that the notion of eco-cultural security, which is of central relevance to indigenous peoples, can also contribute to the revisioning of sovereignty.

Paul Wapner takes up this suggestion in chapter 11, arguing that the state's absolute sovereignty over the environment within its territory, enunciated in the nineteenth century, has evolved in the late twentieth century into "restricted sovereignty," which incorporates the general principle of "good neighborliness." Principle 21 of the Stockholm Declaration articulates this tension between the state's sovereign right to exploit its own resources and its responsibility not to harm the environment beyond its borders. The "rights" side of the equation, however, is much easier to understand and practice than the "responsibility" side. Consequently, the international community has devised numerous mechanisms, most notably liability regimes and regulatory regimes, in an attempt to institutionalize global responsibility. A more radical approach is The Hague Declaration's proposal for a supranational environmental institution to be granted the authority to make policy in the absence of unanimity. Yet Wapner argues that the greening of sovereignty could proceed in more incremental ways—for instance, by involving transnational NGOs to a greater extent in regime formation and implementation. This would tend to scramble the process of environmental diplomacy across state/society lines and across national borders, contributing to the deterritorialization of environmental politics.

Daniel Deudney outlines a far-reaching rendering of this scrambling and resurrects the notion of popular sovereignty in the final chapter of the book. Deudney argues that sovereignty should not be abandoned by environmentalism, but rather that world political theory should "reuse, repair, and recycle" the concept of sovereignty. The key questions would be, who is sovereign, and which patterns of legitimate authority and communal identity are consistent with a particular sovereign, especially given the need to incorporate an intergenerational time frame with respect to environmental issues. Deudney draws on John Dewey's notion of the public, or all those who are significantly affected by the indirect consequences of transactions, to build his notion of an intergenerational sovereign public. He goes on to argue that planetary *topophilia*, or a strong feeling of belonging to the earth, could meaningfully fuse scientific understandings and religious reverence into a communal identity that would serve as the basis for global village sovereignty.

One theme pervades all of the chapters: Sovereignty is on the move, catalyzed at least in part by environmental factors. It is not necessarily being "eroded" in the face of ecological exigencies; nor is it, taken as a whole, being bolstered in any straightforward way. At a minimum, states' autonomy is circumscribed increasingly by international environmental agreements, even as their capabilities are extended into new arenas. On a more fundamental level, some of the chapters suggest that new patterns of control, autonomy, and authority are being spurred by international environmental responses, and that these pose a potentially radical challenge to conventional understandings of sovereignty. The chapters in this book attest to the view that sovereignty is a cluster of practices undergoing multiple processes of unbundling, contestation, and reconfiguration. As "environment" increasingly inserts itself in the world politics of the twenty-first century, we can expect the scope, the complexity, and the intensity of those processes to grow. The greening of sovereignty is a relatively new dynamic in world politics, and while some trends may be discerned, its ultimate trajectory is by no means certain.

Notes

1. I have developed some of the ideas contained in this chapter, particularly the notion of sovereignty bargains, in an earlier essay. See Karen T. Litfin, "Sovereignty in World Ecopolitics," *Mershon International Studies Review* 41 (1997): 167–204.

2. On the transformation of sovereignty, see J. Samuel Barkin and Bruce Cronin, "The State and the Nation: Changing Norms and Rules of Sovereignty in International Relations," *International Organization* 48, no. 1 (Winter 1994), pp. 107–30; and Ruth Lapidoth, "Sovereignty in Transition," *Journal of International Affairs* 45, no. 2 (Winter 1992), pp. 325-45.

3. Richard Falk, for instance, envisioned a benevolent, participatory form of supranational authority, whereas William Ophuls foresaw the emergence of a rather authoritarian and coercive world government. See Falk, *This Endangered Planet: Prospects and Proposals for Human Survival* (New York: Vintage Books, 1971), and Ophuls, *Ecology and the Politics of Scarcity* (San Francisco: W.H. Freeman, 1977).

4. This position is often expressed in articles published in *The Ecologist*. For a critique of both theses, that sovereignty is being (or should be) eroded from above or from below, see Andrew Hurrell and Benedict Kingsbury, eds., *The International Politics of the Environment* (Oxford: Clarendon Press, 1992): 6–9.

5. R.B.J. Walker, *Inside/Outside: International Relations as Political Theory* (New York: Cambridge University Press, 1993): 1.

6. See Jens Bartelson, *A Genealogy of Sovereignty* (Oxford: Cambridge University Press, 1995); Thomas J. Bierstecker and Cynthia Weber, eds., *State Sovereignty as Social Construct* (Oxford: Cambridge University Press, 1996); Janice E. Thomson, "State Sovereignty and International Relations: Bridging the Gap between Theory and Empirical Research," *International Studies Quarterly* 39 (Summer 1995); R.B.J. Walker and Saul H. Mendlovitz, eds., *Contending Sovereignties: Redefining Political Community* (Boulder, Colo.: Lynne Rienner, 1990).

7. I explore this argument in K. Litfin, "Ecoregimes: Playing Tug of War with the Nation-State System," in Ronnie D. Lipschutz and Ken Conca, eds., *The State and Social Power in Global Environmental Politics* (New York: Columbia University Press, 1993): 94–117.

8. An example of this is the Convention on the International Trade in Endangered Species (CITES), which seeks to assist developing countries through nongovernmental organizations (NGOs) in policing and monitoring the ecosystems of endangered species.

9. Gabriela Boyer and Nancy Watzman, "GATTastrophe," *The Nation* 258 (June 13, 1994): 821.

10. Ken Conca, "Rethinking the Ecology-Sovereignty Debate," *Millennium* 23, no. 3 (Autumn 1994): 703.

11. On social construction in international relations, see Alexander Wendt, "Anarchy Is What States Make of It," *International Organization* 46, no. 2 (Spring 1992): 391–425; and Nicholas Onuf, *World of Our Making* (Columbia: University of South Carolina, 1989).

12. For a more extensive discussion in the various schools of international relations thought, and their relationship to environmental concerns, see Litfin, "Sovereignty in World Ecopolitics," 167–204.

13. F.H. Hinsley, *Sovereignty* (New York: Basic Books, 1966): 1.

14. Alan James, *Sovereign Statehood: The Basis of International Society* (London: Allen and Unwin, 1986). See also Lapidoth, "Sovereignty in Transition." Nicholas Onuf has labeled this definition "legalistic and anachronistic." See his "Sovereignty: Outline of a Conceptual History," *Alternatives* 16 (1991): 430.

15. It should be noted that just as property constitutes a bundle of rights that can be disaggregated into use rights, hereditary rights, and so on, so too does sovereignty. Indeed, sovereignty and private property have much in

common, and contemporary environmental exigencies seem to be reconfiguring both.

16. Hinsley, *Sovereignty*.

17. Thomson, "State Sovereignty in International Relations," 223, 230.

18. Max Weber, *The Theory of Social and Economic Organization* (New York: Free Press, 1964).

19. See Barkin and Cronin, "State and the Nation," and Lapidoth, "Sovereignty in Transition."

20. Robert H. Jackson, *Quasi-States: Sovereignty, International Relations, and the Third World* (Cambridge: Cambridge University Press, 1990).

21. Neorealists conceptualize sovereignty in terms of "self-help," or the state's decisional autonomy. For interdependence theorists, however, the state's decreasing ability to cope with permeable borders entails the "erosion" of its sovereignty. Thus, at least part of the divergence between the conclusions of neorealists and those of interdependence theorists stems from the fact that they highlight different dimensions of sovereignty. See Kenneth N. Waltz, *Theory of International Politics* (Reading, Mass.: Addison-Wesley, 1979), and Mark Zacher, "The Decaying Pillars of the Westphalian System," in James Rosenau and Ernst-Otto Czempiel, eds., *Governance without Government: Order and Change in World Politics* (Cambridge: Cambridge University Press, 1991).

22. Edmund S. Morgan, *Inventing the People: The Rise of Popular Sovereignty in England and America* (New York: Norton, 1988).

23. Janice E. Thomson, *Mercenaries, Pirates and Sovereigns: State-Building and Extraterritorial Violence in Early Modern Europe* (Princeton, N.J.: Princeton University Press, 1994).

24. Mark A. Levy, Robert O. Keohane, and Peter M. Haas, ed., *Institutions for the Earth: Sources of Effective International Environmental Protection* (Cambridge, Mass.: MIT Press, 1993).

25. Peter M. Haas with Jan Sundgren, "Evolving International Environmental Law: Changing Practices of National Sovereignty," in Nazli Choucri, ed., *Global Accord: Environmental Challenges and International Responses* (Cambridge, Mass.: MIT Press, 1993): 419.

26. Bruce Byers, "Ecoregions, State Sovereignty, and Conflict," *Bulletin of Peace Proposals* 22, no. 1 (1991): 72.

27. See, for instance, Paul Wapner, *Environmental Activism and World Civic Politics* (Albany: State University of New York Press, 1996); Thomas Princen, Matthias Finger, and Jack Manno, "Nongovernmental Organizations in World Environmental Politics," *International Environmental Affairs* 7, no. 1 (Winter 1995): 42–58.; Ronnie D. Lipschutz with Judith Mayer, *Global Civil Society and Global Environmental Governance: The Politics of Nature from Place to Planet* (Albany: State University of New York Press, 1997).

28. Hugh Dyer, "EcoCultures: Global Culture in the Age of Ecology," *Millennium: Journal of International Studies* 22, no. 3 (Autumn 1993): 483–504.

29. Friedrich Kratochwil, "Of Systems, Boundaries, and Territoriality: An Inquiry into the Formation of the State System," *World Politics* 39, no. 1 (1986), 29.

30. John Gerard Ruggie, "Territoriality and Beyond: Problematizing Modernity in International Relations," *International Organization* 47, no. 1 (Winter 1993), 144–73. Ruggie suggests that, since territorial exclusivity is emblematic of modernity, then the unbundling of territoriality would be a productive avenue of study for exploring systemic transformation.

31. For a sampling of indigenous perspectives on nature, see Interpress Service, *Story Earth* (San Francisco: Mercury House, 1994).

32. William C. Clark, "Environmental Imperialism?," *Environment* 35, no. 7 (September 1993): 1.

33. J. Brooke, "Brazil Announces Plan to Protect the Amazon," *New York Times*, 1 April 1989; quoted in William Wood, George Demko, and Phyllis Mofson, "Ecopolitics in the Global Greenhouse," *Environment* 31, no. 7 (September 1989), p. 17.

34. Caroline Thomas, *New States, Sovereignty, and Intervention* (Aldershot, Hauts: Gower, 1985): 146–47.

35. From the *Beijing Declaration of 41 Developing Countries*, paragraph 6, quoted in Hurrell and Kingsbury, *International Politics of the Environment*, 45.

36. Conca, "Rethinking the Ecology-Sovereignty Debate," 707–10; Joel Migdal, *Strong Societies and Weak States* (Princeton, N.J.: Princeton University Press, 1988).

37. Nancy Lee Peluso, "Coercing Conservation: The Politics of State Resource Control," in Lipschutz and Conca, *The State and Social Power*, 46–70.

38. Byers, "Ecoregions, State Sovereignty and Conflict," 69–70.

39. Robert Livernash, "The Growing Influence of NGOs in the Developing World," *Environment* 34, no. 5 (June 1992): 15, 20.

40. The conflation of authority with coercive power, so prevalent in the international relations literature, would seem odd to most other social scientists. The state's authority is generally perceived as resting not in coercion, but in the citizens' belief that obedience is somehow in their interest. On the relationship between power, authority, and legitimacy, see M. Weber, *Theory of Social and Economic Organization*.

41. Warren Magnusson, "The Reification of Political Community," in Walker and Mendlovitz, *Contending Sovereignties*, 52.

42. Consider the collection of essays in *Institutions for the Earth*, the aim of which is to demonstrate that institutions matter. The presence of international institutions (operationalized as international organizations) made a difference in virtually every case, but ultimately the only reason they made

a difference was because of popular pressure. Levy et al., *Institutions for the Earth*.

43. For some cultural readings of sovereignty, see Ashis Nandy, "The Politics of Secularism and the Recovery of Religious Tolerance," and R.B.J. Walker, "Sovereignty, Identity, Community: Reflections on the Horizons of Contemporary Political Practice," in Walker and Mendlovitz, *Contending Sovereignties*, 125–44 and 159–85; and Christine DiStefano, *Configurations of Masculinity: A Feminist Perspective on Modern Political Theory* (Ithaca, N.Y.: Cornell University Press, 1991).

I

Theoretical Tensions

2

Between Sovereignty and Environment: An Exploration of the Discourse of Government

Thom Kuehls

The starting point of international relations is the existence of *states* . . . each of which . . . asserts sovereignty in relation to a particular portion of the earth's surface and a particular segment of the human population.[1]

In most texts on international politics, the statement above is the extent of the discussion concerning the relationship between sovereignty and environment.[2] Put simply, states assert sovereignty over their environment; that is what makes them sovereign states. And while there may be brief discussions of the legal or strategic aspects of how sovereignty is asserted over an environment, rarely do we come across a discussion of the environmental characteristics necessary for the creation of a sovereign state.

This should come as no surprise to the majority of students of international politics. After all, if sovereignty is thought of as effective rule making, or having the final say, the characteristics of the environment over which sovereignty is asserted seem rather inconsequential. I believe it to be no stretch to assert that this understanding of sovereignty's relationship to environment holds sway in present-day thought concerning international politics, even within the growing field of environmental international politics. However, this is a rather uncontextual, or ahistorical, understanding of the practice of sovereignty.

In the late twentieth century, with the rise of "global environmental crises," environment has been argued to pose a serious problem for sovereignty. In the past twenty to thirty years there has been an explosion of texts dealing with the problematic relationship between environment and sovereignty. For the most part, these texts have focused on the penetrability of sovereign state borders. What has been convincingly demonstrated is that certain aspects of environment are not easily contained within or prevented from entering these sovereign spaces.

Given what could be called *environmental flows* (human and animal travel, waterways, ocean currents, air currents, currency and capital exchanges, etc.), the ability of sovereignty to effectively environ an environment has been called into question. The United Nations' World Commission on Environment and Development, in a widely influential report titled *Our Common Future*, put the problem this way: "The physical effects of our decisions spill across national frontiers."[3] When physical effects of decisions made elsewhere manifest themselves in your sovereign space, your ability to exercise sovereignty has been problematized.

The recognition of the penetrability of sovereign spaces has been the most prominent aspect of the problematic relationship between environment and sovereignty raised over the past thirty years. It is an aspect raised by several of the essays in this collection. It is not, I should emphasize, an unimportant aspect of this problematic relationship. However, when this penetrability thesis has been raised it typically has not brought with it an understanding of the complexity of sovereignty's relationship to environment. In other words, it has not brought with it an understanding of the role government, or *governmentality*,[4] plays in this relationship.

For Jean-Jacques Rousseau, sovereignty is about the rule of law; government, however, is about "political economy."[5] Government involves questions about the character of the environment: How it is that a political community interacts with its environment. What are the people and the land like? How can they be shaped in order to bring about a particular type of community? Where must the political community be compromised to adhere to the environment? In order to establish a particular type of political community known as the *modern sovereign state*, Michel Foucault, drawing on Rousseau's work, maintains that "the formation of a whole series of specific governmental apparatuses" becomes necessary.[6] Creating a sovereign state demands engaging the art of government to appropriately shape the environment. In other words, sovereignty, as it is linked to the existence of the modern state,[7] requires a particular type of citizen and a particular orientation to the land. If we ask the question, what is the relationship between environment and sovereignty?, we are implicitly raising the question of government. What I am suggesting is a need to explore, historically, practically, and materially, how it is that the sovereign state came into existence and continues to exist.[8]

What I propose to do in this essay, through an exploration of Rousseau's discourse on government, is to draw out another way in which environment and sovereignty are linked, beyond environment being simply a space over which sovereignty is asserted. In order to accomplish this task I take several steps. First of all, I present Kenneth Waltz's popular reading of Rousseau as a structural international relations theorist, or realist, developing Waltz's concept of the sovereign state along the way. Second, I present a reading of Rousseau's theory on politics focusing on his discourse on government that cuts against Waltz's reading and ultimately the conception of the sovereign state Waltz articulates. I am utilizing Waltz's works for two reasons: First of all, while his general theoretical perspective has been thoroughly critiqued over the past two decades,[9] his reading of Rousseau continues to carry a great deal of weight in the arena of the history of international political thought, and since I draw heavily from the work of Rousseau to make my argument, it may be worthwhile to "rescue" Rousseau, as it were, from Waltz; second of all, I believe that the general conception of the sovereign state Waltz articulates is still widely accepted throughout the field, and as such, it is illustrative of the ahistorical understanding of sovereignty and the sovereign state that pervades the discipline.

In the third section of this essay I analyze an eighteenth-century document from Portuguese America drawing out the governmental practices at work in constructing sovereign territory. In the last section I return to the World Commission on Environment and Development's proposed solution to our current environmental crises. First of all, I argue that the governmentality put forth by this leading articulation of the "penetrability thesis" is all too similar to the governmental practices involved in constructing the sovereign state examined in sections two and three. Last, I suggest the need for a different governmentality, one that would seriously problematize our current construction of sovereign space.

Waltz Reads Rousseau

The work of Jean-Jacques Rousseau occupies a prominent position in the tradition of international political thought. Perhaps the most popular interpretation of Rousseau as international political theorist has come from Kenneth Waltz.[10] In *Man, the State, and War: A Theoretical Analysis*, Waltz declares Rousseau's writings to be invaluable in isolating the causes of war. "Rousseau's explanation of the origin of war among states is," according to Waltz, "in broad outline, the final one so long as we operate within a nation-state system."[11] Rousseau's explanation is granted the final word in Waltz's text because: "Rousseau himself finds the major causes of war neither in men, nor in states, but in the state system itself."[12]

For Waltz, the most important moment in all of Rousseau's writings is to be found in *A Discourse on the Origin of Inequality*, in the allegory of the stag hunt. Describing the condition of "man" in the state of nature, prior to the evolution of civil society, Rousseau relates this thought: "If it was a matter of hunting a deer, everyone well realized that he must remain faithfully at his post; but if a hare happened to pass within the reach of one of them, we cannot but doubt that he would have gone off in pursuit of it without scruple and, having caught his own prey, he would have cared very little about having caused his companions to lose theirs."[13]

"The story is simple;" says Waltz, "the implications are tremendous." What Waltz reads into the above sentence from Rousseau (which is prefaced by a description of "men" at a certain stage of development prior to the formation of political entities, where they have "some crude idea of mutual commitments" but beyond that no "foresight whatever")[14] is a structural analysis of the inherent problems of interacting units in an anarchic system. In a system marked by anarchy, Waltz hears Rousseau telling him that the potential for conflict among the actors is always present. Cooperation can never be assured. The difficulty of the stag hunt, Waltz concludes, lies not in the particular characteristics of the hunters—their lack of foresight, for instance—but in the structure of their relationship. This example, Waltz contends, "contains the basis for [Rousseau's] explanation of conflict in international relations."[15]

What for Rousseau is an example of one historical stage (prior to the formation of political communities) of the long evolution of human characteristics and societies, becomes for Waltz the paradigmatic description of "man's" political problem. We are self-interested, present-oriented beings by nature, and things are in somewhat short supply, Waltz seems to be saying through this reading of Rousseau. The allegory of the stag hunt is timeless for Waltz. The men in Rousseau's allegory come to represent all human beings, in all times and places. The characteristics of these men are not examined, nor are the characteristics of the forest they inhabit. States, the international arena, and environmental problems easily slide in Waltz's international political theory to replace Rousseau's men, their relationship to one another, and the animals of the forest. With no overarching authority to ensure cooperation, contemporary states must behave in a self-interested manner in order to survive, Waltz informs us."[16] Hence, state sovereignty proves troubling to attempts at resolving transnational environmental problems. But Rousseau's political analysis offers us another approach to these environmental problems.

For Rousseau, as I read him, "man" is far from a static entity. A primary element of *A Discourse on the Origin of Inequality* is the argument that "man" changes throughout history with changes in social structures and territorial features. One of Rousseau's central criticisms of philosophers who came before him is that they took "man" out of a particular context and then held that "man" up as natural and universal. Politics in this analytic mind-set becomes a problem of uncovering rules to fit this (fabricated) natural "man." The political problem for Rousseau, on the other hand, involves both shaping "man's" nature to fit the type of community you wish to create and shaping your political community to meet environmental constraints. (For my purposes in this essay, *environment* includes human and nonhuman elements.) Rousseau is well aware that human beings are not plastic to be molded to fit any frame. Not all humans and peoples can be shaped into citizens or legitimate political communities.[17] The importance of all this is that Rousseau aims to write about a very "contextual" politics. Characteristics are crucial to Rousseau, both of the people who inhabit political communities and the territory upon which they live.

This aspect of Rousseau's work is absent both from Waltz's analysis of it and, more importantly, from Waltz's own political theory. In order to construct his theory of international politics, Waltz claims he "must leave aside, or abstract from the characteristics of the units."[18] The units Waltz needs for his structural theory of international politics must be devoid of character. He declares that he must leave aside "questions about... social and economic institutions... [and] the cultural, economic, political, and military interactions of states."[19] In short, Waltz refuses to ask the very questions that Rousseau is intent on asking: questions concerning environment and questions about government. To ask these questions would confuse the issue at hand for Waltz. For us to think about international politics, Waltz insists, the sovereign state must be ahistorical, abstract, and characterless. But for Rousseau, the sovereign state is none of these things.

Rescuing Rousseau

Rousseau's thoughts on sovereignty, legitimacy, and government pose a series of damaging questions to Waltz's reading of Rousseau's work, as well as to the conception of the state Waltz utilizes. Where Rousseau takes us when we begin to examine the intricacies of his thought is to an understanding of the sovereign, territorial state as a complex construct, not just theoretically, but historically and materially. It is this aspect of Rousseau's thought that I find particularly pertinent to the question of the relationship between environment and sovereignty.

"It suffices to read in chronological succession," argues Foucault, "two different texts by Rousseau": *A Discourse on Political Economy* and *Of the Social Contract*.[20] In the former, Rousseau sets out the problem of government, or political economy, as a problem of political practice. In the latter, he attempts to apply this principle of government within the juridical confines of a principle of sovereignty. The problematic of government, Foucault argues, exploded as a general political problematic in the sixteenth and seventeenth centuries. Newly emerging nation-states began to be concerned with the general welfare and character of the population. This problematic of government brought into the space of the modern sovereign state an entire range of problems from "wealth, resources, means of subsistence, the territory with its specific qualities, climate, irrigation, fertility, etc.... [to] customs, habits, ways of acting and thinking, etc."[21]—things Rousseau maintained were unknown to the political *theorists* of his day.[22]

What political theorists had become obsessed with was the question of *legitimacy*—but a particular version of legitimacy, one that (seemingly) sought to leave environmental questions aside. Two shifts seem to have taken place sometime in the seventeenth and eighteenth centuries in Western political thought. First of all, arguments were being made that all "men" were generally equal to one another—not just in an abstract manner, but in a real and material manner, with respect to emotion, intellect, bodily functions, etc. These arguments connected to a second shift to a belief that there was only one natural, reasonable, or legitimate type of political system. Particularly with respect to human populations, the argument seemed to be that since "we" are all similar by nature, there ought to be only one correct type of political system for all peoples. Moreover, rather than attempting to shape human beings to fit particular political communities, this new political community, known perhaps best as the *liberal democratic sovereign state*, had been made to fit us, as we "naturally" are. A reading of some of the leading voices of this political thought with an eye toward Rousseau's discourse of government, however, will draw out the extent to which the "natural man" of modern liberal democratic thought is, in fact, an explicit creation.[23] And as Foucault has compellingly demonstrated, this construction is not confined to the realm of theory, but is about actual political practice.[24]

What makes Rousseau's work distinct in this context is not his abdication of a search for the "natural man," but his *explicit* recognition that politics cannot realistically be about this "natural man." "Man" in political society is a work of art; government is the artificer. In *A Discourse on Political Economy* Rousseau sets himself to the task of developing the principles of government, that is, how it is that sovereignty and environment are linked. Government, he declares, is about "everything required by the locality, the climate, soil, moral customs, neighborhood, and all the particular relationships of the people ... an infinity of details, of policy and *economy*."[25] Government involves all kinds of problems thought inconsequential to the concept of the sovereign state by Waltzian theorists of politics. It is not enough for government to wield a "public" force to control the private uses of force; government must also tackle an infinity of details of policy and economy that extend beyond the scope of law and order. In short, government must come to terms with the characteristics of environment.

It is important to keep in mind that Rousseau is not theorizing an abstract, ahistorical state. He is, rather, involved in the process of nation building, of state construction. This process involves much more than the presence of a de facto sovereign. In fact, Rousseau explicitly rejects such a power as sovereign.[26] While a sovereign state cannot be reduced to a de facto sovereign for Rousseau, what the state may be has much to do with the environment upon which it might be set. Again, as I am using the term *environment*, it involves both the human and nonhuman elements of an area. Rousseau deals with both.

"If it is good to know how to employ men as they are," Rousseau says, "it is far better to make them such as one needs them to be. The most absolute authority is that which penetrates to the interior of the man and exerts itself no less on his will than on his actions. *It is certain that the peoples in the long run are what the government makes them to be.*"[27] Again, for Rousseau, government is an active, productive force that deals with a broad range of problems. While it is the task of government to "make peoples," Rousseau does not see human material as putty in the hands of governmental practices. The power of government is limited, but that is not to say that it is ineffective.

How then does government make people? "[T]o form citizens is not the affair of a day," Rousseau informs us, "and to have them as men, one must instruct them as children."[28] The enforcement of law and the imposition of punishment is not sufficient toward this end. Government is about morals, opinions, customs; an attention to an infinity of details. Human beings do not come ready-made to be citizens of a state; they must be made fit for sovereignty. This is not a simple task, as is evident by Rousseau's claim: "He who dares to undertake the instituting of a people, ought to feel himself capable, as it were, of changing human nature."[29] But it is not just human nature that they government must be concerned with. The territory must also be of primary interest.

"Having spoken of the general *economy* in relation to the government of persons, it remains for us to consider it," Rousseau continues, "in relation to the administration of goods."[30] Here we are in the realm of what today is commonly considered to be *economics*. Rousseau appropriately calls it *political economy*, for the question of how a society administers goods among its population is a political question, whether that society chooses to adopt a "free-market" system or not. The task of political economy is not just how a society administers goods among its population, but how it appropriates or produces these goods. In raising this issue, Rousseau raises questions concerning how a society should relate to the land upon which it resides. If "[i]t is not enough to have citizens and to protect them; [but] it is also necessary to think of their subsistence,"[31] then government must also concern itself with the territory.

Rousseau devotes book 3 chapter 8 of *Of the Social Contract* to a discussion of climate, land, production, and surplus. He prefaces this discussion with the claim that "the civil State can subsist only as long as the labor of the men produce more than their own needs."[32] This is not an argument for exports. It is, rather, an argument for security; it is necessary to have a reserve to provide for the population in case of lean years, accidents, etc. In a book about establishing principles of political legitimacy, Rousseau speaks at length about issues of production. He is making, here, a direct link between sovereignty and environment. States require surplus if the welfare of the population is to be provided for. To produce this surplus, government must take into consideration the characteristics of the territory.

Where Waltz claims to avoid these issues almost completely in his theory of international politics for the purpose of being able to assume a (characterless) unified actor (known as the *sovereign state*), Hedley Bull's theory of international politics provides a skeletal description of the necessary characteristics of the sovereign state. In *The Anarchical Society* Bull argues that in order for a political community to be a state it must "possess a *government* and assert *sovereignty* in relation to *a particular portion of the earth's surface* and *a particular portion of the human population*."[33] As Bull takes us through a brief look at political communities that do not constitute states under this definition, he ends up excluding a wide range of nonmodern, non-European political communities. Statehood is denied to the peoples of "Africa, Australia and Oceana, before the European intrusion" because "there was no such institution as government" present at the time of contact.[34] Since Bull feels that these communities had no government, he sees no reason that they should have been dealt with according to the rules of international relations. It appears that being accepted as a state is not just a question of having a sufficient amount of force to make other states recognize you, for certainly numerous states exist today and have existed over the centuries that have not had this level of force yet have been dealt with as such. To be a state, and hence treated as a sovereign entity, requires not only sovereignty, but population, territory, and something Bull calls *government*.

It is not my intention to engage in a sleight of hand with respect to the political terminology I am utilizing. In no way am I suggesting that government means the same thing for Bull as it does for Rousseau. For Bull, government appears to be no more than an institution. He does not seem to be granting it the productive aspects that Rousseau did. Nonetheless, it is a term worth exploring briefly, given the larger argument I am attempting to make here.

What is this thing Bull calls *government*, that these nonmodern, non-European societies are lacking? Vine Deloria, Jr. and Clifford Lytle speak to this issue in *The Nations Within: The Past and Future of American Indian Sovereignty*.[35] According to Deloria and Lytle, American Indian nations have been denied sovereignty for a range of reasons that could be argued to fit into Bull's use of government theory. In particular, Indian societies did not seem to contain an organized structure of political authority.

> To the Europeans, Indians appeared as the lowest form of man. No formal institutions were apparent. Leaders seemed to come and go almost whimsically. One might be negotiating with one chief on one occasion and be faced with a different person for no apparent reason...In tracing the source of political authority, whites were really baffled. No one seemed to be in charge of anything.[36]

Rather than seeing this "lack" of political organization as evidence of a less developed political culture, Gilles Deleuze and Felix Guattari suggest that we think of it as a direct attempt to ward off the state-form.[37] In other words, the state is a particular type of political organization. It is not necessarily about the ability to defend borders from the inside or outside, but is rather about being organized in a particular manner. And that manner of organization goes beyond having a particular foundation for political authority; it also involves how a society is organized to interact with the land upon which it resides.

Governmentality and the Construction of Sovereignty in Portuguese America

In 1757, the governor of Portuguese America issued the *Diretorio dos Indios*. This directorate granted the native Americans residing within Portuguese-American colonial territory full legal rights and called for the transferal of "the temporal authority of the missionaries [over the natives] to secular administrators."[38] The stated purpose of this document was to civilize, or make "citizens" out of, the natives. The process of making citizens was not an end in itself. The American natives were to be transformed into citizens so that they would become "useful to themselves" and "to the State."[39] Their use to the state would be exhibited primarily through their taking part in transforming the jungle into more economically useful land forms, thereby contributing not only to the general wealth of the state, but to the sovereignty of the state over its environment. The process of creating citizens did not involve simply granting rights to preexisting "subjects"; rather, it required a *creation* of "subjects" capable of bearing rights within a sovereign state structure. A reading of the directorate's passages reveals the political technologies applied to both people, in order to make them citizens, and land, in order to make it sovereign territory.

To begin with, the "Indian" was to be transformed into a European citizen through an attack on "Indian" culture. Traditional dress was to be replaced with European-style clothes—especially among women. "Indian" social groups were to be restructured into individual family units, in part through the construction of European-style cottages with separate rooms. Town jails were constructed. Two yearly lists were to be created: The first would have names of persons who had diligently worked the land; the second would have names of persons who preferred to live in idleness[40]—which, when read in opposition to "diligently worked the land," could mean virtually anything else. While the roles of clothes, family structure, architecture, and penal practices are crucial to understanding the creation of citizens and sovereign states, the practices I focus on here involve the land more directly.

The director to be placed on "Indian lands" was "to teach [the 'Indians'], not so much how to govern themselves in a civilised way, but rather how to trade and cultivate their lands . . . [which would] . . . make these hitherto wretched people into Christians, rich and civilised."[41] The civilizing process would be carried out primarily through the process of transforming "Indian lands" into Portuguese farms. To be made into citizens, these "Indians" would have to be convinced of the "dignity of manual labor."[42] Through their labors they would be transformed; they would become citizens. By working to transform the land from jungle to farmland or harvestable forest this "wretched people" would become a "civilized people."

Economic activity, it seems, according to this directorate, is primarily what makes someone a citizen—not to mention a Christian. But a specific type of economic activity is called for in this directorate, one that carries with it a particular orientation to the land, one that will provide the surplus necessary to constitute a sovereign state. For example, the "casual" method of "harvesting the forest" which was used by the "Indians" was to be replaced by a more "organized" form "so that only the most lucrative commodities [would be] gathered."[43] The question, Lucrative for whom, or what?, immediately comes to my mind. The shift in farming practices and methods of harvesting the forest was undertaken in order to establish, in the words of the directorate, "an opulent and completely happy state."[44] Where "Indian" economic activity had been oriented toward providing for the immediate community, it would now have to be oriented toward the state. This would require two transformations: one in the "Indians" and one in the land.

The "Indians" had used the forest for centuries before the Portuguese invaded. But their methods of "harvesting the forest," their political economy, was found to be lacking by modern European standards. In fact, the dichotomy between diligently working the land and living in idleness suggests that for the authors of this document "Indian" methods of using the land did not count as work but fell into the category of idleness. And in order for Portuguese America to become a state, the people's characteristics had to change so that the characteristics of the land could change. If the transformation of Portuguese America into a state was to be successful, the "Indian" would have to be transformed from a communal, undisciplined hunter-gatherer, into a single-family, disciplined, productive unit of the state.

This is a problem of sovereignty. The point I am trying to make here is that the governmental practices put in place through this directorate are bound up with an attempt to "sovereignize" Portuguese America.[45] The attempt by Portuguese-American authorities to turn the jungle into property is directly related to a claim of sovereignty over what was to become Portuguese-American territory. An absence of property signals an absence of sovereignty. And to hold a property in the land, one must relate to that land in a particular way. The land must be coded. The "smoothness" of it must be eradicated through the use of the forces of striation.

John Locke's theory of property is operative in the Portuguese America directorate. Locke argued, in the late 1600s, that all "men" have a natural right to property through their labor. And labor, for Locke, as in the Portuguese America directorate, was a specific activity. Drawing his connection between labor and property, Locke left us with this description of wasteland: "Land that is left wholly to nature that hath no improvement of pasturage, tillage, or planting, is called, as indeed it is, waste."[46] On Locke's reading, property only came into being when land had been subjected to particular types of use. All other methods of procuring the fruits of the earth are captured in the phrase "left wholly to nature." Thus, a society that did not put plow to earth, or domesticate and herd cattle, but instead grew crops in a manner that produced far less visible signs of land use (to the European) and acquired meat by hunting animals that had free roam of the land, could not be said to have a property in that land; and consequently, could not be said to hold sovereignty over that land. Thus the nations of Europe were able to justify their expansion across the surface of the earth on the grounds that they were filling vacant, or *smooth*, spaces.

In his nineteenth-century examination of democracy in America, Alexis de Tocqueville reached similar conclusions in his discussion of the condition of the North American territory and the relation of the "Indians" to it: "Although the vast country that I have been describing was inhabited by many indigenous tribes, it may justly be said, at the time of its discovery by Europeans, to have formed one great desert."[47] As did Locke, Tocqueville maintained that the indigenous, nonmodern peoples who inhabited the North American expanses could not be said to hold any claim to the land because they did not utilize it in an appropriate manner. For these people to become civilized, Tocqueville continued, and hence, for them to constitute a sovereign state, the "wandering propensities" of the "Indians" had to be checked. They had to be "settled permanently."[48] Sovereignty requires a particular orientation to the land.

Once that orientation has been demonstrated, vacant spaces and wandering propensities drop from sight. For instance, Locke allowed that a political community of property holders may consent to leave some lands "in common" and "unused" and not run the risk of having "outsiders" lay claim to this land through their labor.[49] Similarly, Tocqueville often speaks of the wanderlust of the Europeans who were continually packing up and heading further west across the American continent, yet somehow their wandering propensities did nothing to detract from their being "civilized," or members of a sovereign state. For unlike their Indian neighbors, the wandering Europeans brought forces of striation with them. Even in Locke's case of the land left "in common," it is striated, coded, and made to fit into the overall scheme of the sovereign state.

Drawing on the thought of Gilles Deleuze and Felix Guattari in *A Thousand Plateaus*, one can see the project of sovereignty as one of striation—a refusal to allow the existence of "smooth space."[50] America constituted a smooth space, a desert for Tocqueville, in large part because it lacked signs of striation, the signs of sovereignty. It was populated, in Tocqueville's eyes, by nomads, not civilized peoples. But is the nomad to be understood solely in regard to his or her lack of permanence in a particular place? Is this what gives the state its sovereignty and denies the same to the nomad?

"The nomad has a territory," Deleuze and Guattari contend.[51] And even Tocqueville demonstrates some recognition of this claim, for he not only chastises the "Indians" for their "wandering propensities," but for their "instinctive love of country [that] attaches them to the soil that gave them birth."[52] Indeed, it is the Europeans who are the ones constantly on the move, according to Tocqueville, moving first across the Atlantic Ocean and then continually further west across the American continent. In the face of this expansion, the "Indians" are now criticized for refusing to move off the land of their birth. What then makes them nomads?

It was not whether or not the "Indians" *were* attached to a particular tract of land that mattered for Tocqueville, but *how* they attached themselves to this land. "The Indians occupied [the land] without possessing it."[53] Europeans, on the other hand, seemed to "possess" whatever lands upon which they stopped. The difference lies in the orientation to the land. Where the orientation of sovereignty *"parcels out a closed space to people,* assigning each person a share and regulating the communication between shares, the nomadic trajectory . . . *distributes people (or animals) in an open space,* one that is indefinite and noncommunicating."[54] The sovereign trajectory operates in and with striated space, while the nomadic trajectory operates in and with smooth space. The two require different orientations to the land. As such, the two require different types of people.

Our Common Future and the Globalization of Governmentality

In this section I return to the claim put forth by the World Commission on Environment and Development (WCED) in *Our Common Future* that sovereign territorial borders are penetrated by the environmental effects of decisions made elsewhere. More specifically, I turn to the solution proposed by the WCED for this situation: sustainable development. Last, I argue that if our current environmental problems stem from the governmentality described in this chapter, then perhaps what is needed is a different type of governmentality and I offer some possibilities here.

The "penetrability thesis," as put forth by the WCED, goes somewhat as follows: Sovereignty has been problematized in part because of transnational environmental problems. No state is safe from this globalizing crisis if for no other reason than "the physical effects of our decisions spill across national frontiers."[55] In the face of this reality, the WCED maintains that traditional notions of security no longer hold. The belief that states can isolate themselves from this situation and hide behind their well-defended territorial borders is a false one. "We are now forced to accustom ourselves to an accelerating ecological interdependence among nations. Ecology and economy are becoming ever more interwoven—locally, regionally, nationally, and globally—into a seamless net of causes and effects."[56] We face a new world, the WCED seems to be saying, one where state borders are more like permeable membranes than sovereign walls. The environmental crisis is intensely local in some respects and remarkably global in others. It penetrates all states. The solution to this crisis, the WCED seems to be saying, must operate along the same lines.

Recognizing the problem, the WCED seeks to identify both the cause and the solution. Over several years, the WCED held public hearings on five continents around the globe. Through these hearings "one central theme" is said to have emerged: "Many present development trends leave increasing numbers of people poor and vulnerable, while at the same time degrading the environment."[57] Such a conclusion might well have taken the WCED down a path similar to the one I have sketched out in this essay; a path that would have led the members to examine the practice of government in the creation of sovereign states; a path that would have led them to examine how environments have been shaped to fit into a sovereign state mold. However, this is not the path the WCED set out on after uncovering this "one central theme."

In a section in Chapter 1 titled "Symptoms and Causes," the WCED seems to distance its argument from the "one central theme" when it declares that "poverty itself pollutes the environment."[58] The possibility that poverty itself may be a consequence of development projects, or that development projects may pollute, has disappeared as the WCED now sets the stage for declaring "sustainable development" to be the hope for the future. Since poverty pollutes and development's professed goal is to eradicate poverty, development can now be cast in a "green" light. Granted, the WCED does maintain that sustainable development will not be development as usual. Recognizing that development without resources is impossible, the WCED makes a case for a new globalizing development strategy that will do a better job of conserving the world's resources for future development. The primary goal of the WCED, after all, is to devise a plan to sustain development, not necessarily to protect the environment in all its complexity. The problem of sustaining development is identified as a transnational one, and hence, the argument goes, it will require a transnational solution. But does this solution seriously challenge the prevailing orientation to the land sketched out in the Portuguese America directorate, Locke's theory of property, and Tocqueville's discussion of the fate of the American Indians? Or does it merely seek to keep this orientation alive by arguing for a more efficient use of resources?

What I find troubling about the WCED's proposed solution is its failure to engage the problem of government. Recognizing the problem posed by environmental interdependence is not enough if the governmentality of the solution remains the same. In other words, if the governmental practices involved in the construction of the sovereign state analyzed in this chapter are at all contributing factors to the global environmental crises we now face, simply globalizing these governmental practices hardly seems like a solution.

Our Common Future is long on generalities and short on specific programs. Yet an analysis of a project endorsed there may help to shed light on the governmentality that is operating in this version of sustainable development. "One important initiative that attempts to put conservation more squarely on the agenda of international development," the WCED informs us, "has been the Tropical Forestry Action Plan."[59] Issued by the United Nations Food and Agriculture Organization (UN-FAO), along with the World Resources Institute and the World Bank, the Tropical Forestry Action Plan (TFAP) makes a case for industrial forestry as a way of saving trees. Claiming that it is "the rural poor themselves who are the primary agents of destruction as they clear forests for agricultural land, fuel wood, and other necessities,"[60] the TFAP goes on to argue that these tropical forests must be taken out of the hands of the "rural poor" and placed in the hands of global forestry industry. As products for export, these forests will be better cared for and will be able to make an important "contribution to the well-being of nations in the developing world,"[61] as they simultaneously make an important contribution to the global economy.

The governmentality of the TFAP is not far from the governmentality of the eighteenth-century directorate from the governor of Portuguese America. Both seek to transform the environment to mesh with an all-too-similar political economy. Both seek to striate the forests, rather than to leave them "wholly to nature." According to Indian environmental activist Vandana Shiva, projects such as those outlined in the TFAP are "aimed at changing the character of the forests in such a manner that they will exclusively serve commercial interests, and not the indigenous peoples."[62] Tree farms are placed in areas deemed to be "wasteland," but wasteland, Shiva adds, is an economic, not an ecological, category. The diversity of life that occupied these acres of tropical forest is removed in order that, as one international forestry consultant put it, "trees and parts of trees of preferred species . . . can be profitably marketed." From this perspective, most of the species of this tropical biomass are "clearly weeds."[63] What smooth character of the forest did exist is to be overcoded under this Tropical Forestry Action Plan.

If this is sustainable development as the WCED understands it, then I cannot imagine how it solves the problem of environmental destruction. If the particular governmental practices that lie between sovereignty and environment are the cause (something the WCED's "one central theme" might suggest), then all the WCED's penetrability thesis accomplishes is a globalization of these practices. If we continue to see lands that are not generating revenue for the state or the global economy as waste-lands, and if we continue to ignore the human inhabitants of these lands by declaring them to be vacant, I cannot see how we can come to grips with the environment/sovereignty problematique. What is needed, I am suggesting, is a different governmentality.

Looking over the other chapters in this collection, I can see at least two possibilities for exploring different governmentalities: One option lies in "ecosystem management," as discussed in Veronica Ward's Chapter 4; another lies in Franke Wilmer's efforts to "take indigenous critiques seriously (Chapter 3)."[64] A shift to an ecosystem management philosophy where ecology takes precedence over economics would have a dramatic effect on how governmental practices would shape environments. Sustainable development, as it is presented by the WCED, does not do this. Economics still takes precedence over ecology; development is what is to be sustained, not ecosystems. Depending on how the boundaries of ecosystems would be drawn, an ecosystem management approach could have the effect of challenging our conception of sovereign space on two levels: geopolitically and governmentally. Given current lines of sovereignty, it may be impossible to put an ecosystem management policy into effect, for many ecosystems cross national borders and many effects on ecosystems may come from several states away. Moreover, a true ecosystem management program would radically alter current political economic systems which are part and parcel of the construction of geopolitical space. Because of this potentially doubly unsettling effect on sovereignty, it should come as no surprise that it is in Antarctica where ecosystem management is getting its first shot, a land claimed by no state and offering little revenue-generating capacity for the global economy.[65]

Similarly, a real shift in governmental practices would occur were nonmodern perspectives taken seriously and lands inhabited by non-moderns not deemed *vacant*. There is often a different political spatiality in nonmodern societies. As I noted above, drawing on the work of Deleuze and Guattari, where the sovereign state orientation to political space is based on lines of striation, nonmodern orientations often require an acceptance of smooth space. Moreover, an acceptance of nonmodern perspectives would no longer require the same transformations in humans (and lands) to become fit subjects within sovereign states. To "take indigenous critiques seriously" might shift us, in our thinking, away from that damaging dichotomy between "natural" and "under human use." Nonmodern peoples do not necessarily leave the lands they live upon "wholly to nature," as Locke's simple dichotomy might suggest. In many cases there is a sophisticated management of natural resources going on in nonmodern political economies.[66] Operating in "our" particular governmentality, "we" are often unable to see "their" use for the forest.

Both "taking indigenous critiques seriously" and following an ecosystems management approach would have a significant impact on the sovereignty/environment relationship because of the shift in governmental practices they would require. The modern sovereign state is a particular political construction for which environments do not come ready-made. The task of molding environments to fit the sovereign state form is that of government. To ask a question about the relationship between sovereignty and the environment is to enter into an analysis of the discourse of government; it is to enter into an analysis of how environments are shaped to coincide with the parameters of that particular modern political community known as the *sovereign state*.

Notes

1. Hedley Bull, *The Anarchical Society* (New York: Columbia University Press, 1977): 8.
2. By *environment* I mean not simply the territory and any plants and nonhuman animals that may inhabit it, but the human population as well.

3. World Commission on Environment and Development, *Our Common Future* (New York: Oxford University Press, 1990): 27.

4. The term *governmentality* is drawn from Michel Foucault's lecture of the same name, in G. Burchell, C. Gordon, and P. Miller, eds., *The Foucault Effect* (London: Harvester Wheatsheaf, 1991): 87–104.

5. Jean-Jacques Rousseau, "A Discourse on Political Economy," trans. C. Sherover, in *Of the Social Contract and Discourse on Political Economy* (New York: Harper and Row, 1984): 143. Henceforth, Rousseau's "Discourse on Political Economy" will be refered to as Rousseau, *DPE*.

6. Foucault, "Governmentality," 103. For Foucault, governmental apparatuses are not limited to the halls of sovereignty, or what we might normally understand to be "the government." Foucault has in mind a broader sense of the term, which draws in the various disciplinary practices that operate to govern us, from medical discourse, to education, to religion, etc., yet which are not necessarily in the hands of "the government." In this sense his use of *governmentality* exceeds Rousseau's use of *government*.

7. See F. H. Hinsley, *Sovereignty* (New York: Cambridge University Press, 1986), for a discussion of the link between the sovereignty and the state.

8. This is not, obviously, a new issue in international political thought. Theorists ranging from John G. Ruggie to R. B. J. Walker have launched critiques of mainstream international political theory for its ahistorical conception of the state. See, for instance, Ruggie, "Continuity and Transformation in the World Polity: Toward a Neorealist Synthesis," *World Politics* 35, no. 2 (1985), and Walker, *Inside/Outside: International Relations as Political Theory* (New York: Cambridge University Press, 1993).

9. The most powerful critique of Waltz's theoretical project, to my mind, comes from several essays by Richard Ashley, particularly "The Poverty of Neorealism," *International Organization* 38, no. 2 (1984); and "Living on Border Lines: Man, Poststructuralism, and War," in J. Der Derian and M. Shapiro, eds., *International/Intertextual Relations: Postmodern Readings of World Politics* (Lexington, Mass.: Lexington Books, 1989).

10. Kenneth Waltz, *Man, the State, and War* (New York: Columbia University Press, 1959). For other "(neo)realist" readings of Rousseau's work, see M. Forsythe et al., eds., *The Theory of International Relations: Selected Texts from Gentili to Treitschke* (London: Allen and Unwin, 1970); and I. Clark, *Reform and Resistance in the International Order* (New York: Cambridge University Press, 1980). While I am aware of compelling critiques of the (neo)realist reading of Rousseau's works from authors such as Stanley Hoffman ("Rousseau on War and Peace" in *The State of War* [New York: Praeger, 1963]) and Michael Williams ("Rousseau, Realism, and 'Realpolitik,'" *Millennium* 18, no. 2 [1989]), I will not discuss them here, for neither builds his reading of Rousseau's works on Rousseau's discourse of government.

11. Waltz, *Man, the State, and War*, 231.

12. Ibid., 6.

13. J.-J. Rousseau, *A Discourse on Inequality*, trans. M. Cranston (New York: Penguin Books, 1984): 111 (henceforth referred to as Rousseau, *DOI*).

14. Rousseau, *DOI*, 111.

15. Waltz, *Man, the State, and War,* 167.

16. According to Waltz, while the logic of a global Leviathan to control the self-interests of the various states may be unassailable, this remedy is "unattainable in practice" (ibid., 238), in part, because of the level of violence that will be necessary to bring a global state into existence. Waltz focuses on a police violence, a negative violence, a prohibitory violence, but the use of violence can be extended to a productive violence, a government violence that is always (also) necessary to sovereignize an environment.

17. See Rousseau's discussion in *Of the Social Contract*, in Sherover, *Of the Social Contract and Discourse on Political Economy* (henceforth referred to as Rousseau, *SC*) book 2, chapter 8, of the inability of some peoples ever to achieve freedom.

18. Ibid., 79.

19. Ibid., 80.

20. Foucault, "Governmentality," 101.

21. Ibid., 93.

22. Rousseau, *SC*, 51.

23. William Connolly has provided an excellent reading of Hobbes, Rousseau, Sade, Hegel, and Marx along these lines in *Political Theory and Modernity* (New York: Basil Blackwell, 1988).

24. See, for starters, Foucault's "Two Lectures" in C. Gordon, ed., *Power/ Knowledge* (New York: Pantheon, 1980), where Foucault urges us to "try to discover how it is that subjects are gradually, progressively, really, and materially constituted through a multiplicity of organisms, forces, energies, materials, desires, thought, etc.," 97. See also *Discipline and Punish*, trans. A. Sheridan (New York: Vintage Books, 1979); and *History of Sexuality Volume One: An Introduction*, trans. R. Hurley (New York: Vintage Books, 1980), both by Foucault. Again, Foucault thinks of politics or government on a wider scale than we might normally think of them. For Foucault, and for myself as well, in this essay, the practices of government exceed the confines of the juridical state apparatus.

25. Rousseau, *DPE*, 150, emphasis in original.

26. Rousseau, *SC*, 7.

27. Rousseau, *DPE*, 151, emphasis mine.

28. Ibid., 160.

29. Rousseau, *SC*, 37.

30. Rousseau, *DPE*, 164.

31. Ibid.

32. Rousseau, *SC*, 74.

33. Bull, *The Anarchical Society*, 8.

34. Ibid., 9.

35. Vine Deloria, Jr., and Clifford Lytle, *The Nations Within* (New York: Pantheon Books, 1984).

36. Ibid., 9.

37. Gilles Deleuze and Felix Guattari, *A Thousand Plateaus* (Minneapolis: University of Minnesota Press, 1987): 357.

38. C. Machlachlan, "The Indian Directorate: Forced Acculturation in Portuguese America (1757–1799)," *The Americas* 28, no. 4 (1972): 359.

39. J. Hemming, *Amazon Frontier* (London: Macmillan, 1987): 14.

40. This list of political technologies from the directorate is drawn from both Machlachlan, *The Americas*, and Hemming, *Amazon Frontier*.

41. Quoted in Hemming, *Amazon Frontier*, 11–12.

42. Quoted in Machlachlan, *The Americas*, 363.

43. Ibid., 365.

44. Quoted in Hemming, *Amazon Frontier*, 16.

45. The attempt continues in present-day Brazil. Witness the government's policy of encouraging expansion west into the Amazon rain forest to fill an area deemed "empty" by the Brazilian state despite the presence of thousands of indigenous inhabitants. Under the Plan for National Integration, begun in 1960 with the intent of "occupying" the Amazon rain forest as fast as possible, the state provided incentives for miners, ranchers, and other settlers to enter this area. Official estimates are that the population of the state of Amazonia in Brazil increased from 100,000 in 1960 to 2 million in 1970 (A. Diegues, "Social Dynamics of Deforestation in the Brazilian Amazon: An Overview" [Geneva: UN Research Institute for Social Development, 1992]: 10). The occupation of Amazonia serves to sovereignize the territory, demonstrating visible signs of "use," as it brings it further into the fold of the Brazilian (and world) economy.

46. John Locke, *Two Treatises of Government*, ed. P. Laslett (New York: Mentor, 1965): 339.

47. Alexis de Tocqueville, *Democracy in America*, vol. 1 (New York: Vintage Books, 1945): 26.

48. Ibid., 356. For a further discussion of this general reading of the North American Indians by Europeans and its effects see Deloria and Lytle, *The Nations Within*.

49. Locke, *Two Treatises of Government*, 334.

50. Deleuze and Guattari, *A Thousand Plateaus*; in particular see "plateaus" 1, 12, 13, and 14.

51. Ibid., 380.

52. Tocqueville, *Democracy in America,* 351.

53. Ibid., 26.

54. Deleuze and Guattari, *A Thousand Plateaus,* 386.

55. WCED, *Our Common Future,* 27.

56. Ibid., 5.

57. Ibid., 4.

58. Ibid., 28.

59. Ibid., 161.

60. Quoted in V. Shiva, "Forestry Myths and the World Bank," *The Ecologist* 17, no. 4-5 (1987): 142.

61. *Tropical Forestry Action Plan* (Rose: United Nations Food and Agriculture Organization, 1985): 85.

62. Shiva, "Forestry Myths and the World Bank," 142.

63. Quoted in J. Bandyopadhyay and V. Shiva, "Chipko: Rekindling India's Forest Culture," *The Ecologist* 17, no. 1 (1987): 33.

64. See F. Wilmer, "Taking Indigenous Critiques Seriously: The Enemy 'R' Us," and V. Ward, "Sovereignty and Ecosystem Management: Clash of Concepts and Boundaries?," chapters 3 and 4 in this collection.

65. In this respect Ward's chapter and mine are somewhat at odds. She maintains, "It is not, however, the fact or existence of sovereignty itself, in whatever guise or form, that is responsible for the absence of such a program. It is the absence of commitment to the principles that define the ecological perspective" (draft, p. 20). I would respond that Ward is thinking of sovereignty in an unproblematic relationship to environment. She is not thinking of sovereignty's relationship to environment as one influenced by the practice of government.

66. For a further discussion of this issue, see S. Hecht and A. Cockburn, *The Fate of the Forest: Developers, Destroyers, and Defenders* (New York: Harper and Row, 1990): chapters 2, 8, and 9.

3

Taking Indigenous Critiques Seriously: the Enemy 'R' Us

Franke Wilmer

The first thing that makes the west different from every other place is the overwhelming and compulsory theory and practice of Progress, the ideology of Progress, the most powerful ideology in the world, by which people are urged to believe that with the accumulation of technical information, and scientific information, and written arts, and the spread of the world economy as we know it, that things are getting better and better and better. I think we can kind of date the Theory of Progress to about 1680... sometime around John Locke's time. That's also the time, by the way—the 1680s—we notice that there is the beginning of the disappearance of species and all kinds of animals in the world. So for 300 years while things have been getting better, and better, and better in the world view of the west, lots of things have been disappearing from the face of the earth, so that for at least some species, things have been getting worse, and worse, and worse.

—John Mohawk (1992)

Within the past several decades indigenous peoples have become increasingly active in international fora.[1] Their political message is straightforward: They wish to be recognized as sovereign peoples insofar as this entails acknowledging their right to live in continuity with their own continually evolving cultural communities and to control their own political destinies. On numerous occasions, however, they have also imparted a second message that speaks to the heart of the intersecting topics to which this volume is addressed—sovereignty and the natural environment. From Nunavut to Yucatan, indigenous leaders warn that *human action and its impact in the world are directed by a view that is dangerously out of touch with natural laws which, according to indigenous peoples, govern all life on this planet.* These "natural laws" require that human activity be balanced and limited by acknowledging interdependence among diverse life forms within interconnected ecosystems.

The limits of the natural environment flow from its status not only as an object, but as a *subject with its own capacity for agency.*

To view the natural environment as a subject with a capacity for agency contrasts sharply with a view of the natural environment as an object or a "thing" to be acted on, manipulated, and controlled, or in the language of the Enlightenment, to be tortured and conquered. But my claim is not that indigenous views of the environment treat nature as a subject while Western views treat it as an object. Nature/resources-as-object is always implicated in economic exploitation, and indigenous economies certainly always did and do now exploit the natural world in this way. It is rather that indigenous views suggest a more complex relationship to the environment as *both* object and subject, with important consequences for how we understand the ability (and limitations) of humans to act in relation to the environment. Sovereignty in practice has meant conceiving of the state as a subject whose capacity to act on objects is unrestrained. Once nature is assigned any subject status from the perspective of human interpretation, this means that certain processes of the natural world are beyond human control, and so beyond *sovereignty.*

The views expressed by indigenous peoples in a variety of international, regional, national, and local forums constitute a powerful critique of the dominant, Western-originated worldview. In the tone of an imperative, indigenous delegations spoke to a special session of the United Nations titled Cry of the Earth. The Haudenosaunee or Six Nations of Iroquois called their Address to the Western World in Geneva in 1977 "A Basic Call to Consciousness."[2]

This essay takes indigenous critiques seriously, including the contention that, notwithstanding important differences and variety among indigenous perspectives, there is a common underlying epistemological distinction between indigenous and Western thinking as a whole which is particularly salient on the issue of the human-nature relationship. After outlining its main elements with illustrative examples from indigenous addresses to international audiences, I will examine the differences in Western and indigenous thought on sovereignty and the environment by suggesting that the origins of these differences lie in the way in which self-Other[3] relations are constructed within the two perspectives.

Western epistemology constructs the environment almost exclusively as an object, and therefore as a repository for its own projective identification. The natural world is alternately viewed as the nurturant Other to be controlled, on the one hand, or the problematic (chaotic, hostile, threatening, even "evil") self projected as a malevolent Other. To understand more fully how this cognitive move takes place, I employ feminist psychoanalytic accounts of the construction of the Western self. This suggests that at least part of the problem can be traced to the way in which masculinity and femininity reflect self-Other constructions, and the ways these constructions are conceptualized and socially transmitted within different cultural contexts.

Indigenous Knowledge/Indigenous Voices

What if indigenous peoples are *not* "primitive," *not* socially, culturally, philosophically "less evolved" than Western peoples?[4] What if different cultures simply construct different versions of reality, premised on different values, priorities, methods of acquiring knowledge, and different conceptions of what constitutes knowledge? In the view of Vine Deloria, Jr. "tribal peoples are as systematic and philosophical as Western scientists in their efforts to understand the world around them. They simply use other kinds of data and have goals other than determining the mechanical functioning of things." Indigenous peoples' knowledge of the ecosystems in which they live, for example, is the product of observations and experiences transmitted orally through generations extending to the earliest human habitation of an area.[5] What if there are simply different paths of cultural evolution, each with its own production of knowledge, and, perhaps most importantly, each culturally specific form of knowledge containing a potentiality for contributing to the solution of our common, global problems? A collective statement by the Native American Council of New York City put it this way: "When whole cultures are lost, so are different ways of thinking, distinct perspectives and philosophies that allow humans and nature to live in harmony. At the same time as the world is rapidly being deprived of its biological diversity and its ability to sustain itself, it is also being robbed of the tools of thought that may counter this self-destruction."[6]

By attending to interpretive, culturally specific, and historically contextualized construction of knowledge, knowledge is revealed as information acquired by culturally determined methodologies, interpreted within culturally determined worldviews, and applied to culturally defined problems. *Technology* can be thought of as the application of (socially constructed, culturally and historically situated) knowledge. This opens the door to the possibility for a myriad of technologies to be developed according to socioculturally constructed questions, methodologies, and (historically and culturally contextualized) conceptions of the most pressing problems. It also opens up important new spaces for cross-cultural discourses and knowledge sharing.

Indigenous Critiques of Western Ways of Thinking

Western cultures—for the past millennium or so, but particularly since the Enlightenment—have been on a trajectory focused on the development of material technologies arising out of the ways in which Western thinkers have interpreted their own experiences and problems. Furthermore, much of the thrust of this technology is directed toward a natural environment conceived of as hostile, chaotic, and unpredictable. Not only is the natural environment conceived of in this way, but so too is the field of international relations itself. Anarchy and chaos are seen as the problem; sovereignty, the extention of political control, is the solution.

Although postmodern, postcolonial, feminist, and ecofeminist writers have pointed us in the direction of critically examining how the West conceives of the Other in relation to gender, non-Western peoples, and the natural and political environments, little attention has been paid to alternative cultural constructions of self-Other relations.[7] Beyond the parameters of critical discourses—anthropology as cultural critique, for example, and some indigenous academics[8]—indigenous knowledge systems are dismissed out of hand as naïve at best, and "primitive" at worst. Yet there is absolutely no basis for rejecting indigenous knowledge systems other than the purely ideological project of establishing and maintaining a socially constructed reality that will support the continued material and philosophical dominance of Western elites in the world.

Many indigenous activists believe that asserting a right to sovereignty as conceived within Western thought, and usually within the parameters of self-determination, is necessary in order to limit state actions aimed at their cultural assimilation and consequent cultural destruction.[9] But a somewhat graver message is never wholly separated from these claims. It is a call not only to recognize, through self-determination, indigenous peoples' right to allocate resources as a means of perpetuating their own cultural practices and adaptations, but is directed as a more general admonition to and critique of the *excesses* of Western materialism. According to indigenous critics, the Western world is distinguished by its failure to maintain a balanced relationship between human activities and the natural environment. Materialistic excess drives Western notions of "development" to destroy indigenous peoples and their cultures, which are then cast as an obstacle to "development." Virtually all other complaints and criticisms presented in public arenas by indigenous representatives—greed; arrogance; paternalism; expropriation of in- digenous peoples from their land through forced relocation, coerced assimilation, ethnocide, and genocide—can be traced to this overriding theme.

It is not that Western culture alone is materialistic and exploits resources while indigenous peoples are not and do not.[10] Those indigen- ous peoples who resist wholesale cultural assimilation (hereinafter referred to simply as *indigenous* enjoy a material existence, and ac- knowledge that all human economic activity exploits resources.[11] None advocate the abandonment of Western material technology.

Change in the lives of indigenous peoples is a condition which has always existed. Serious changes have given rise to serious readaptations to the new condition. Indigenous peoples represent many peoples, many cultures, and also different ways of thinking. But they share the same natural world and the same spiritual world...we cannot help but make the observation that industrial political states have risen and seem to be in decline since their emergence just over two hundred years ago. Tribal societies have existed for over 10,000 years and continue to adapt and adjust. Which is the better way, growth and consumption or balance?[12]

Western technologies are commonly blended with preexisting tradi- tional indigenous knowledge by indigenous peoples themselves. But by focusing on the creation of surplus value *without moral restraint*, Western societies appear to indigenous peoples to be exceedingly, even stupidly and self-destructively, greedy. The problem is not with capital- ism (or the technologies it has produced) but with the fact that in itself it does not constitute a *moral system* and, in fact, seems to displace local cultures and human value systems. As Haudenosaunee writer Sottsisowah[13] writes: "It is a hard fact of life that the misery which exists in the world will be manipulated in the interests of profit.... The roots of a future world which promises misery, poverty, starvation, and chaos lie in the processes which control and destroy the locally specific cultures of the peoples of the world."[14]

Indigenous critics call for a fundamental shift in the orientation of dominant economic and political decision makers toward the natural world, claiming that such a shift is necessary to the collective future of humanity. The following excerpts from public addresses by indigenous activists, representing many different indigenous groups primarily to international audiences, are illustrative:

We are here due to the urgency of the cry of our Mother Earth and the urgency of our concern for all forms of life, physical and spiritual.... [w]e have a choice for the future...but only if we listen and join again the spiritual worlds. Otherwise the balance of the earth will shift her axis and life will be greatly destroyed and harmed by this process.[15]

The global environmental crisis has more than adequately demonstrated that business as usual will not and cannot ensure global survival. What is needed is a fundamental shift in consciousness...[16]

Brothers and Sisters: We cannot adequately express our feelings of horror and repulsion as we view the policies of industry and government in North America which threaten to destroy all life. Our forefathers predicted that the European Way of Life would bring a Spiritual imbalance to the world.... Now it is before all the world to see—that the life-producing forces are being reversed, and that the life potential is leaving this land.[17]

The indigenous peoples...expressed their points of view throughout the entire UNCED process.... With only changing our attitude a little and by investing some effort in protecting the earth through peaceful cooperation in living, nature can regenerate itself to the benefit of all.... This change in attitude can affect how we view the exploitation of natural resources with the resultant contamination, degradation, and accumulation of riches in a few hands.[18]

We must realize for once and for all that the destruction of indigenous peoples of the world is also the destruction of mankind.... Among the obstacles to our well-being is that which they call "development" or "progress," which has mistakenly led to unwise state policies...even deforming our vision of the world and the cultural identity of each people.[19]

We come together because we are alarmed by the destruction of vital life structures.... Water is primary to life; corn is next. Poisoned water will poison all life; lack of water causes droughts, deserts and death. The nations that sit in the great council of the United Nations must relearn the natural law and govern themselves accordingly, or face the consequences of their actions.[20]

It may not be coincidental that the destruction of indigenous peoples occurs in the same period in which it is a necessity that the whole world change the way it is interacting with the environment.... We need a rebirth of thought. A rebirth of thought which focuses the attention of the west from where it is now—which I have to tell you it is now focused on oblivion—to the whole issue of what is the human potential for survival in the coming natural hard times ... hard times that are a direct result of human agency directed at nature.[21]

The views of indigenous individuals excerpted here are not intended to suggest that there is a single, coherent, uncontested, monolithic "indigenous" or "pan-Indian" view, nor am I suggesting that these speakers represent "the" indigenous view. Rather, these speakers have both self-identified and are widely regarded within their respective local communities as thoughtful people whose perspective is shaped by their historical and cultural experience *as indigenous thinkers*, in contrast to "Western" thinkers. That indigenous peoples have had, and many maintain, a very different orientation toward the environment is not news to Westerners. The stereotypical "environmental Indian" is by now well known, popularized, and even commercialized in the media, although rarely considered beyond its appropriation as a trivialized cliché or a form of objectification that reinforces the dichotomous, often romanticized, image of primitive natives compared to advanced moderns.[22] But I will attempt to push beyond the stereotype of "indigenous peoples living (harmoniously) in a state of nature" to propose that the indigenous critique goes to the heart of the cognitive, ideological, and philosophical issues that undergird what some see as modernity's slide into a potential environmental Armageddon. Indigenous critiques claim, ultimately, that the project of indigenous peoples saving themselves is intimately linked to changing the way nonindigenous peoples think about themselves and the world, as indicated by many of the excerpts quoted above. From this perspective, the link between our conception of sovereignty and our environmental practices is absolutely central to the necessity of cognitively restructuring our politics.

Several caveats should be kept in mind when considering what is distinctive about indigenous philosophies when contrasted with Western philosophies. The first is the obvious problem of totalizing either or both. Western and indigenous historical experiences and discourses are multifarious, internally contentious, and multivocal. However, it is possible, I believe, to talk about common strands of thought within cultural traditions as long as one keeps in mind that cultural traditions are neither static nor monolithic, but rather continually adapting and engaged not only in intercultural critiques but in intracultural debates. They are both traditions and discourses, and are continually adapting to the contemporaneous forces of intercultural exchange.

The second caveat is that when one is comparing and contrasting two or more culturally grounded philosophies, the analyst must acknowledge the limitations imposed by her own position within one of those cultures. As a woman of European descent socialized in the Western world, I cannot claim that my analysis is free of my own projections— perhaps all perceptions are in part tainted by a projective process. But I attempt to remain conscious of and apply something I have learned from years of listening to indigenous peoples—that Others are *both* objects (at least as far as our potential for projecting ourselves on Others) and subjects. To see Others as subjects in their own right means acknowledging the Other as occupying a position I cannot occupy. An Other-as-subject is an Other with a right to its own unique experience, which can be communicated across selves, but which maintains the integrity of the boundaries of distinctiveness.

Indigenous Philosophies

Political states view uncontrolled growth and progress as the highest ideals, while indigenous groups regard balance and limited growth as essential to their livelihood. From all appearances these ideas cannot be reconciled.[23]

The views of indigenous peoples have sometimes been dismissed as "archaic" or "primitive," meaning that they do not "count" in the "modern" world. This attitude derives not from any falsifiable hypothesis or refutable argument, but rather from the need of Westerners to sustain a belief in the legitimacy of their own materially, politically, and philosophically privileged position in the world: Western civilization is presumed to be the universally desirable path. This belief is predicated on a linear conception of time and cultural adaptation that equates the path of Western cultural evolution with progress. A variety of terms have been used over the past few centuries to denote non-Western peoples' inferiority: "primitive," "backward," "undeveloped," "underdeveloped," and, most recently, "traditional" have come to stand for a way of life that opposes "progress."

As Vine Deloria, Jr., has argued in *The Metaphysics of Modern Existence, God Is Red*, and elsewhere,[24]

The very essence of Western European identity involves the assumption that time proceeds in a linear fashion; further it assumes that at a particular point in the unraveling of its sequence, the peoples of Western Europe became the guardians of the world. The same ideology that sparked the Crusades, the Age of Exploration, the Age of Imperialism, and the recent crusade against Communism all involve the affirmation that time is peculiarly related to the destiny of the people of Western Europe. And later, of course, the United States.[25]

Much has been written elsewhere on the nature of indigenous philosophies.[26] Here I will limit the discussion to several recurring themes that may be most relevant to the issues of sovereignty and the environment.

The first is the idea of an interdependent and interconnected world. Any social construction that interpolates the human-environment relationship, including "sovereignty," occurs within the limits of what is understood a priori as the fact of interconnectedness and interdependence. "Destroy nature and you destroy yourself," said Chumash Juanita Centeno.[27] Interconnectedness is implicit in a circular view of reality.[28] As Jenny Leading Cloud (White River Sioux) put it: "We Indians think of the earth and the whole universe as a never-ending circle, and in this circle man is just another animal. The buffalo and the coyote are our brothers, the birds, our cousins. Even the tiniest ant, even a louse, even the smallest flower you can find—they are all relatives."[29]

It is not only interdependence that distinguishes an indigenous view of the environment—international relations has its own theory or paradigm of interdependence—but also the belief that life forms have equal rights to exist, that is, that there is no moral hierarchy among forms of life, and that they exist within an integrated whole. The idea of a circular, interconnected, nonhierarchical relationship among life forms goes far beyond notions of interdependence in social science theory. As Mohawk Jim Ransom, an environmental specialist at Akwesasne, explains:

The issue is one of the difference between a Circle of Life versus a straight line of hierarchy with people at the head of the system and resources commanded or controlled by people. The Circle of Life in which people and resources are interrelated has been broken. . . . An alternative adaptive model is to repair the Circle by returning to an understanding of the traditional ways and applying those to present problems.[30]

This is not to say that indigenous social systems were ever or are now nonhierarchical. However—and this is particularly important in understanding where they stand on the human-environment relationship— indigenous views of the natural world typically do not assume a *moral hierarchy* among life-forms.

If all life-forms, as subjects, are equally entitled to live, how can one life-form rationalize taking another's life in order to continue its own? The proposition that life-forms are equally important does not exist apart from the *observable* reality that life is a cycle—death is neither the opposite of life, nor the negation of life. Rather it is part of the cycle that *includes* birth, reproduction, maturity, and death, all of which contribute to the continuance of life.

Thus it was that while Indians hunted and fished wild game, they made it a rule that unless they were starving and needed food for survival, they would not take the animals and birds until these creatures had enjoyed a full family life and reproduced their kind. Even today when taking eagles, the Apaches restrict the hunt to late summer or autumn to ensure that the eagles have the chance to mate, raise a family and go through the major cycles of life experiences.[31]

The taking of life is not based on a moral rationalization about the superiority of human existence, or the priority of human life over other forms. Moral exclusion, used to rationalize brutality or what we might term *gratuitous violence* against the Other, can be found among indigenous cultures in behavior directed toward real and tangible human enemies, but even there the violence of self/enemy-other relations is subjected to certain moral restraints.[32] For example, among the Crows and other Plains tribes, the hunter-warrior must look into the eyes of the prey or enemy, and life-taking is accompanied by a prayer acknowledging the connection between killer and victim.[33]

By contrast, among Western cultures, violence or manipulation of the natural world is rationalized on the basis of the moral superiority of humans. This rationale is then turned against the natural world as justification for its destruction (via "dominion"). Indeed, most of the brutality inflicted on native peoples by European settlers was either rationalized on the basis of their moral inferiority, or undertaken as a result of the belief that Western peoples could "improve" the moral character of non-Western peoples via various civilization schemes.[34]

In actuality, indigenous worldviews are based on an interpretation of reality which is constituted as an essentially "moral universe," to use Deloria's terminology. It is moral because it is the result of choices that are continuously being made by all participants, and all choices have consequences and are therefore endowed with moral significance.

> The world is constantly creating itself because everything is alive and making choices which determine the future.... The real interest of the old Indians was not to discover the abstract nature of physical reality but rather to find the proper road along which, for the duration of a person's life, individuals were supposed to walk. This colorful image of the road suggests that the universe is a moral universe: There is content to every action, behavior, and belief.... No body of knowledge exists for its own sake outside the moral framework of understanding. We are, in the truest sense possible, creators and co-creators with the higher powers, and what we do has immediate importance of the rest of the universe.[35]

This interconnection is fundamentally spiritual, but, as Deloria points out, all it takes is a substitution of *energy* for the term *spirit* and "we have a modern theory of energy/matter."[36] The idea is also akin to Plato's notion of the ideational reality behind material manifestation: "We believe that all living things are spiritual beings. Spirits can be expressed as energy forms manifested in matter. A blade of grass is an energy form manifested in matter—grass matter. The spirit of the grass is that unseen force which produces the species of grass, and it is manifest to us in the form of real grass."[37] The world imagined by indigenous philosophies is not only interdependent, but at once diverse, integrated, interconnected, cyclical, and moral. Morality, like the democratic truths declared by America's founding philosophers/settlers, is self-evident: It follows from an awareness of the consequences of the actions of one life-form on other forms in a system in which the continuation of life is the ultimate imperative. In such a world, all life-forms are simultaneously subjects and objects.

Understanding the Difference: A Psychocultural Analysis

We seem to be dancing in the night toward a psychological period of even more irresponsible behavior than we have, as though that were possible. It strikes me that where we're at is we need a period of time in which responsible intellectuals seriously question the irresponsible gross materialism of this culture, and its utter inconsistency with reality.[38]

How is it that, "instead of the predatory jungle which the Anglo-Saxon imagination conjures up to analogize life, in which the most powerful swallows up the weak and unprotected," indigenous perspectives understand life "as a tapestry or symphony in which each player has a specific part or role to play"?[39] John Mohawk points to a difference in self-Other orientations toward the natural world in indigenous and Western thought.

The Iroquois lived in the same world that everybody else lived in, they froze with everybody else, they starved with other people, they lived through the same horrors of natural disaster, but I am clear about the ancients. They didn't ask, "Is it real that Nature is an enemy or is it real that Nature is a friend?" They asked, "Are we healthier if we think of Nature as a benevolent being, or are we healthier thinking of Nature as an opponent?" And they concluded that human beings, to have a healthy way of being in the world, have to conclude that Nature is a benevolent other, and have to want to cooperate with Nature. If you can't cooperate with Nature, then you think that Nature is an antagonistic other. They asked the question, not what is real, but how should we think? How should we behave? What kind of world should we create?[40]

Explorations into the possibility of a postobjectifying consciousness and a liberatory science "that enables human beings to recognize nature as a *subject* in its own right" have been undertaken by critical theorists as well as feminists,[41] and both have revealed the link between patriarchy and the domination of nature, which is also at the core of recent ecofeminist thinking.[42] Objectification is also linked to domination, for objectification *enables* domination. Objects are projections of the self (often "bad" or "undisciplined" aspects of the self; other times romanticized and idealized), but do not have rights of their own. Domination—of the natural world, of women, and "other Others" as objects of "sovereign" control—is the logical outcome of a particular (Western) worldview. How, then, have indigenous cultures come to view the natural world, men and women, and Others, as *both* objects *and* subjects? How did they arrive at a human-nature relationship that is not driven by a project of domination?

Feminist and psychoanalytic theorists trace the development of self-Other relations to infancy and early childhood. The self and Other are twinborn. As the boundaries that delineate the self as a distinctly identified being come into existence, so too is the Other constructed. Whether we view other individuals and the world (natural and social) as a basically friendly or hostile place, is largely determined during this early relationship between an infant and its caregiver(s). Many psychoanalysts agree that the process begins sometime in the first six months of life and continues during the preoedipal period until around the age of two.[43]

The mirror stage runs from about six months to two years of age. At some time during this stage the young child recognizes himself (*moi*) in a mirror. ... He is impressed by the image, the wholeness of the image, the appearance of definite boundaries, form, and control seen in the mirror. Yet the wholeness is really an illusion. He is none of these things. Sometimes he even confuses his own image with the image of his mother who, in one of Lacan's versions, is holding him before the mirror.[44]

The self not only develops as a bounded individual, but as one living in a world of contradictory experiences and impulses. Specifically, psychoanalytic theory suggests that the infant-caregiver relationship is fraught with ambivalence and contradiction arising out the infant's experiences of both needs-gratification and needs-frustration, which appear to be wholly controlled by the caregiver(s). The contradictory impulses of "love" and "hate" (or feelings of adoration and rage) arising out of gratification or frustration experiences are initially resolved by the infant by conceiving of two "Others," one good (needs-gratifying) and one bad (needs-frustrating). The infant's world of objects, then, is originally occupied by a good and a bad Other. When the infant or child feels love or hate in response to the caregiver, he or she also supposes that the self, like the Other, is split between a good and a bad self.

Feminists have further refined the psychoanalytic account by specifying that masculine and feminine identity, which can also be traced to this early period of self-differentiation, develop in potentially different ways in relation to the infant's understanding of his or her own gender as well as that of the primary caregivers. Although some, like Nancy Chodorow and Dorothy Dinnerstein, believe that female primary caregiving or "mother-monopolized" child rearing is universal, I would suggest that there are very significant variations within indigenous traditions (including those practiced at and adapted to the present time) that may account for the greater propensity of those socialized within those traditions to endow the natural world with both object and subject status, and to view the relationships between humans and other life-forms in terms of symbiosis rather than domination. First, however, let us consider the implications of the female-monopolized child rearing that has been prevalent within Western culture, particularly since the advent of industrialization.

As we learn that we are bounded selves, we also learn (through socialization and mirroring) that we are "male" or "female" and that Others are identified in these categories as well. Under conditions of female-monopolized caregiving, male infants experience self-Other differentiation, rage and gratification in relation to the Other (the Other being the mother), along with a growing awareness of the difference of the Other, while female infants will experience self-Other differentiation, rage, and gratification toward the Other, and a growing awareness of *the sameness of the Other*. Consequently, for girls the Other is simultaneously distinct or separate from the self, and categorically the same, whereas for boys the Other is distinct and *different*. Possibly, for a girl the growing knowledge that the Other who is the same represents the female child's own destiny as an adult is a source of comfort for the anxiety she feels regarding the Other's apparent control. But boys will not take much comfort in a growing awareness of their own masculinity and their primary caretaker's femininity. It may, in fact, become the basis for lifelong rage against females, which may account for the widespread incidence of patriarchy and misogyny. In order to develop an integrated psyche, the child must first come to see the "good" and "bad" caregiver as one integrated entity with a dual capacity for behaviors leading to both pleasure and frustration for the child. This then makes it possible for the child to understand her or his own self as similarly constructed: a single integrated self with the capacity to *feel* both love and rage toward Others.

When infant and child caregiving is monopolized by females, psychic integration is potentially much more problematic for males. As female identity develops in relation to the "good" and "bad" (m)Other who is also the same, female caregiving enables female children to accomplish more easily the task of integrating the apparent contradiction and probably to develop more easily a capacity for compassion as well. The world is constituted by Others who have the capacity to do harm as well as to care, to be "good" and "bad," and the self, when integrated, is understood in the same way. Male identity is more problematic because the splitting of "good" and "bad" Others can be sustained through the perceived duality of masculine and feminine identity. A split male psyche perceives Others as feminized and either idealized or demonized. Female-monopolized child care thus tends to result in problematic male identity, which is more prone to disintegration or splitting of both the self and Others. Others are "feminized"; the feminine is unconsciously regarded as all-powerful and therefore threatening; and control of the feminine source of needs-gratification/frustration becomes a primary objective. Masculine identity under these conditions must also rely more heavily on "negative femininity." With little or no positive experience of masculinity through sustained interaction with adult males, male children whose world of Others consists almost entirely of women tend to rely more heavily on masculinity identity as "not feminine."

Male anxiety regarding the Other who controls gratification and frustration of his needs and who is also fundamentally and undeniably different can be mitigated in several ways. Optimally, if we assume a world in which individuals are socialized according to a dichotomous conception of gender, the anxiety of the male infant can be mitigated by developing relationships with adult males. "I will never be a woman [same as the all-powerful Other who controls needs gratification and frustration]," he reasons, "but I will grow up to be a powerful and capable man [adult male]." In this role he can imagine himself acquiring some control over his environment, over the forces that provide gratification of his needs, and empowered to make contributions to the needs of family and community. Additionally, male caregiving makes it possible for male infants to accomplish the same move toward psychic integration and compassion, or the ability to "see" the self in Others. In indigenous cultures, for example, praying for and looking into the eyes of one's enemy acknowledges that "enemy" is only one role each of the two is playing, while at another level they are spiritual and moral equals. This reflects a capacity to view Others in more complex terms, to account for the paradox of killing one who is "enemy" but equally human, to view the enemy-Other as both object (enemy) and subject (equally human).

While there has been little research thus far on the correlation between father-child relations and the construction of the male psyche, Scott Coltrane has conducted a study of ninety nonindustrial cultures, virtually all of which were indigenous, in order to examine the relationship between male involvement in child rearing and the status of women.[45] He found a strong correlation between male participation in child rearing and egalitarian gender relations, access of women to public decision making, and female status in general. Within indigenous communities there remain large sectors of people whose family structures and child-rearing practices are continuous with traditional practices, including elaborate systems of kinship within extended families in which male relatives (mostly uncles and grandfathers on the mother's side, but sometimes on the father's side) participate extensively in caregiving and child rearing. Additionally, in the majority of indigenous cultures in the United States, grandparents were and continue to be primary caregivers, a fact that would result in more direct male participation in child rearing than in a family structure in which one (usually the father) or both parents are otherwise primarily engaged in economic activities. In precapitalist indigenous cultures, there is often extensive contact between adult men and preadolescent and adolescent boys. We should expect to find in indigenous societies with greater male involvement in child rearing an inclination for something other than patriarchal social structures, and this is, indeed, the case.[46] The majority, though by no means all, of indigenous cultures are matrilineal or matrifocal or both.[47] Finally, we might also expect that greater male involvement in child-rearing would produce more psychically integrated adults with a greater capacity to accept contradiction and to tolerate diverse gender identity. This is, indeed, a characteristic of many indigenous cultures.[48]

Female-monopolized child rearing may provide an explanation for the development and maintenance of patriarchal social systems.[49] But patriarchy does not seem to be the problem as much as the consequence, as Chodorow notes, of a social order in which men are insufficiently involved in caregiving . Within these terms, I believe that many elements of the Western worldview revealed by an indigenous critique become understandable: power and sovereignty as "control over," a reflection of masculine insecurity in relation to a male's own empowerment and anxiety in relation to a world constructed as a threatening feminine natural environment that would arbitrarily deny one's needs and must therefore be controlled; objectification of the natural world; linear, dichotomous thinking characterized by delusions of superiority; and the apparent inability to accommodate the contradictions between diversity and unity, difference and sameness. This masculine self cannot see himself in Others, as a result of little or no intimate caring between adult males and male children; nor can he perceive connection and interdependence among different, bounded entities.

The Cognitive Roots of Western Ideas of Sovereignty

Sovereignty is a social construction that has developed in relation to Western conceptions of power and political community. Western cultures are also (but by no means exclusively) characteristically patriarchal, particularly in contrast with indigenous cultures. Sovereignty can be viewed as a masculine defense against a hostile world of Others.[50] But sovereignty has a dual function—as a protection against outside threats, and as a means of control over the resources and territory *within*. Claiming sovereignty over the environment precludes granting the environment any subject status of its own. This may not only be an unethical position from the perspective of environmentalists' claims, but an inaccurate one in terms of understanding ecosystem processes in a way that enables us to live as humans within their constraints.

Feminist psychoanalytic theory helps us deconstruct the cognitive roots of the Western sovereignty–environmental destruction link. The story goes something like this: Western culture is founded on an entirely objectified relationship between humans and the environment. Furthermore, the object appears as a hostile and threatening Other, which is at the same time the source of human needs. The solution or response devised by Western culture has been to control an objectified natural world. A feminist psychoanalytic reading of the position of the Western self in relation to the environment-as-Other suggests that sovereignty was invented as a defense again a natural world (the "predatory jungle") conceived as hostile and chaotic.

Since the production of knowledge in Western culture has been conducted from a primarily masculine perspective, then to make sense out of the institutions and norms flowing from Western knowledge systems we need to understand the process and perspective of male cognitive development, then analyze it within the context of Western cultural practices. Male cognitive development entails coming to terms with the difference of the Other. During infancy and early childhood the Other is perceived as all-powerful, the source of all good (or pleasure and gratification) and all evil (or pain, frustration, and rage). In order to develop a concept of the self as equally powerful with the Other, and thus less threatened by difference, a child must interact extensively with same-sex adults. This is not such a problem for women because of the primacy of female caretakers during infancy and early childhood and in most cases well into adolescence. The relative unavailability of adult males to male children in Western culture, however, produces an inability in men to accommodate the ambiguous cognitive position that understands Other as "same but distinct" and, hence, truly equal. As it happens, in indigenous cultures young boys spend much of their time around adult men. In many indigenous cultures, all of a child's adult male relatives are considered fathers and all female relatives are mothers.

Finally, we should note that the imagery of Western culture as the good (civilized) self and the natural world as a hostile (uncivilized) Other is fundamentally dichotomous. There is now fortunately an intellectual movement in the social sciences and in international relations that is generally associated with postmodern and feminist perspectives on politics, to move "beyond" dichotomies: either/or, inside/outside, reason/emotion, nature/culture, private/public, and so on.

Ironically, a comparative psychocultural analysis of Western and indigenous perspectives indicates that it is the dualism of Western worldviews which is primitive, not the interdependence of differences found in indigenous worldviews. Dualism is found in virtually all cultural traditions, in all cosmologies, but it is interpreted very differently in Western and non-Western cultures. Consider, for instance, the Tao in Chinese culture; the "middle way" in Buddhism; the contradictory nature of creation figures such as Coyote, the Raven, or Trickster, among indigenous cultures. Dualism represents the original position of the reflective self, but mature consciousness expands beyond rigid dualism, and thus beyond the necessity of defending a bounded society of loved objects against a hostile and ever-present external enemy. Likewise, mature consciousness moves beyond the need to control the natural world as hostile and withholding Other—in other words, *beyond sovereignty* as conceived in Western terms. The self and Other remain distinct but connected, bounded but interdependent. Dualism is not the problem, but rather the way dualism is interpreted from the perspective of a split Western self. It is the failure to *mature*, cognitively speaking, *beyond dualism* that is at the heart of the problem. Thinking of the world in exclusively dualistic terms is indicative of the infantile cognitive position. Indigenous perspectives, on the other hand, easily accommodate contradictions, such as those between separation and connection, unity and diversity.

This means getting beyond both the worldview preoccupied with the human ability to control resources and people "within" the boundaries of sovereignty, and the employment of control strategies vis-à-vis forces "outside" sovereign boundaries. Sovereignty, as DiStefano argues, was created from a masculinist perspective as a defense against the hostility of the Other. We need to move from defensive separation to bounded but interconnected difference. It is no accident that in the name of (technological) creativity, expanded without restraint and aimed at "controlling" the Other in its entirety, our present "sovereign" institutions are carrying out a program of complete planetary destruction. The foundation of the world economy, and the military system that it supports, is sovereignty as currently conceived.

Conclusion

One way of understanding the significance of indigenous critiques of Western culture's treatment of the environment is to examine the critique from a feminist psychoanalytic perspective. This exercise suggests that the problem lies with the cognitive development of the Western self. The world system, which is based on the abstract concept of sovereignty as the defining characteristic of states, is embedded within a deeply patriarchal worldview and the result of the ideological hegemony of Western culture. Actors within the system, at both international and national levels of discourse, project onto the world the Western self's own unresolved and sharply dichotomous dualism. This dualism is characterized by an intense aversion to difference, a consequent insecurity about boundaries, and a defensive reaction in the form of unrestricted efforts to control all that is perceived as Other—including the "environment" conceived as "natural resources." The Western self seems inordinately reliant on its formulation of the Other as threatening, chaotic, anarchic, and morally inferior.

A rigid and impenetrable sovereignty may arise out of a felt need for defense against such a world. Nature-as-Other is furthermore subjected to projects (development, industrialization, capitalism, settlement, and so on) aimed at bringing it under human domination. A persistently demonized Other also suggests a self that needs to sustain its own idealized image—a self unaccepting of and denying its own capacity to do harm or to engage in "evil" behavior. The Western masculine self at this point seems incapable of accommodating the apparent contradiction between diversity and bounded Otherness on the one hand, and unity, connection, and interdependence with the Other(s). The inability of prevailing conceptions of sovereignty to meet the needs for circumscribing the limits on environmental exploitation is indicative of a very serious, underlying cognitive failure—the failure of the Western self to reach maturity.

Indigenous knowledge systems, on the other hand, may offer an alternative way of constructing the self-Other relationship. "Enemy" is but one of a variety of images assigned to Otherness. The cognitive shift necessary to account for a powerful but beneficent Other, an Other as *both* object *and* subject, requires developing some means of accounting for contradiction, and a will to move Western culture in the direction of integration, rather than dichotomous splitting. According to psychoanalytic theory, cognitive maturity necessitates two moves. The first is understanding that the forces of "good," or pleasure-giving/needs-fulfilling, and "bad," or pain-inducing needs-frustrating, coexist within the Other. Thus, the Other is no longer *either* benevolent or threatening, but capable of both. The second move involves relocating or reflecting the coexistence of these contradictory forces within the self. This is a process of integration, for it requires integrating the concepts of self and Other, and enables the self to begin to enter into more complex relations with the Other as a distinct but connected entity, capable, like the self, of doing "good" or "evil". To the extent that these moves can be made and more or less sustained, the self will no longer be inclined to see the Other dichotomously as either enemy or friend, and may indeed no longer require external enemies in order to maintain a bounded "good" concept of the (sovereign) self. The ability to comprehend the contradiction implicit in conceptions of the Other (both in nature as well as in diverse cultural communities) as both object and subject—the goal of mature narcissism—then becomes possible, and with it the possibility of reframing sovereignty not as a defense against an inherently hostile world, but as a system for allocating responsibility for a multitude of global actors.

Notes

1. Franke Wilmer, *The Indigenous Voice in World Politics: Since Time Immemorial* (Newbury Park, CA: Sage, 1993).
2. Akwesasne Notes, *A Basic Call to Consciousness* (Rooseveltown, N.Y.: Akwesasne Notes, 1978, 1991).
3. I will capitalize "Other" when it is used to refer to the psychological concept of an "object" or "objects" as distinct from and delineated by the boundaries of the self.
4. See George E. Marcus and Michael M. J. Fischer, *Anthropology as Cultural Critique* (Chicago: University of Chicago Press, 1986).
5. Vine Deloria, Jr., 1996.
6. Native American Council of New York City statement in Alexander Ewen, *Voice of Indigenous Peoples* (Santa Fe, N.M.: Clear Light Publishers, 1994): 20.
7. William E. Connolly, *Identity/Difference: Democratic Negotiations of Political Paradox* (Ithaca, N.Y.: Cornell University Press, 1991); Bill Ashcroft, Gareth Griffiths, and Helen Tiffin, eds., *The Post-Colonial Studies Reader* (New York: Routledge, 1995); Ashis Nandy, *The Intimate Enemy: Loss and Recovery of Self under Colonialism* (Oxford: Oxford University Press, 1983); Christine DiStefano, *Configurations of Masculinity: A Feminist Perspective on Modern Political Theory* (Ithaca: Cornell University Press, 1991); and Irene Diamond, *Fertile Ground: Women, Earth and the Limits of Control* (Boston: Beacon Press, 1994).
8. Marcus and Fischer, *Anthropology as Cultural Critique*. See also Vine Deloria, Jr., "If You Think about It You Will See That It Is True," *Noetic Sciences Review* (Autumn 1993), and Deloria 1996.
9. An obvious potential pitfall is that casting their assertions according to the rhetorical practices of a Western-dominated political and legal order will

undermine their efforts to adapt and maintain continuity with their own rhetorical and epistemological practices.

10. There are ongoing debates about the environmental impact of native peoples prior to European contact in the Western hemisphere. I would urge caution in evaluating these arguments, and recommend Vine Deloria's *Red Earth, White Lies: Native Americans and the Myth of Scientific Fact* (New York: Scribner, 1995) for a studied and well-researched native perspective on these theories.

11. I use the qualifier *traditional* to refer broadly to indigenous peoples who understand themselves as existing within a cultural system that is continuous with their own unique traditions. However, this does *not* mean that these traditions are not continually adapting, and doing so at least in part in relation to indigenous-Western interactions. Traditional indigenous peoples are also distinct from those indigenous peoples (usually a sector within indigenous communities) who believe generally that their own cultural traditions should for the most part be abandoned in favor of reproducing Western social, economic, political, and normative structures. Not everyone in particular indigenous communities would necessarily fall within these two camps, nor would there necessarily be agreement as to who falls into which camp other than by self-identification. An excellent study of how these positions affect contemporary indigenous societies is Melissa L. Meyer, *The White Earth Tragedy* (Lincoln: University of Nebraska Press, 1995).

12. Center for World Indigenous Studies, *Alternatives to Development: Environmental Values of Indigenous Peoples.* Environment Workshop, Northwest Regional Conference on the Emerging International Economic Order (Seattle, Wash., March 30, 1979).

13. *Haudenosaunee* is the Native word for the Iroquois confederacy.

14. Notes, *Basic Call to Consciousness*, 115.

15. William Command, Algonquin, addressing the United Nations, November 22, 1993.

16. Ruby Dunstan, Lil'wat, Lytton Band, British Columbia, in David Suzuki and Peter Knudtson, *Wisdom of the Elders: Honoring Sacred Native Visions of Nature* (New York: Bantam Books, 1992: 234.

17. Haudenosauanee Declaration of the Iroquois, in ibid., 240.

18. Jose Dualok Rojas in Noel J. Brown and Pierre Quiblier, eds., *Ethics and Agenda 21: Moral Implications of a Global Consensus* (New York: United Nations Development Program, 1994): 51.

19. Noeli Ocaterra Ulian, Wayuu Indian from Venezuela, speaking to the United Nations, December 10, 1992.

20. Statement of the elders brought by Oren Lyons to the United Nations, August 29, 1982.

21. John Mohawk, "What We Can Learn from Indigenous Cultures," lecture at Hunter College, New York, January 1993, taped and unpublished.

22. Greta Gaard, ed., *Ecofeminism: Women, Animals, Nature* (Philadelphia: Temple University Press, 1993).

23. Paper prepared by the World Council of Indigenous Peoples, National Indian Lutheran Board, National Congress of American Indians, and others for the Northwest Regional Conference on the Emerging International Economic Order, "Alternatives to Development: Environmental Values of Indigenous Peoples," March 30, 1979.

24. Vine Deloria, Jr., *The Metaphysics of Modern Existence* (San Francisco: Harper and Row, 1973); *God is Red: A Native View of Religion* (Golden, Colo.: Fulcrum Publishing, 1994). See also Deloria, "If You Think about It." Deloria, Standing Rock Sioux, has been involved in indigenous political activism for over twenty five years, and has written extensively on indigenous law, philosophy, rights, and politics. He currently teaches in both the law and political science programs at the University of Colorado at Boulder.

25. Deloria, *God Is Red*, 63.

26. Akwesasne Notes, 1978; Deloria, "If You Think about It"; Menno Boldt and J. Anthony Long, eds., *The Quest for Justice: Aboriginal Peoples and Aboriginal Rights* (Toronto: University of Toronto Press, 1985); Suzuki and Knutdson, *Wisdom of the Elders*, 1992.

27. Steve Wall, *Wisdom's Daughters: Conversations with Women Elders of Native America* (New York: Harper-Collins, 1993): 45.

28. Franke Wilmer, "Narratives of Resistance: Postmodernism and Indigenous World Views," *Race, Gender, and Class Special Edition: Domination and Resistance of Native Americans* 3, no. 2 (1996): 35–58.

29. Wilma Mankiller and Michael Wallis, *Mankiller: A Chief and Her People* (New York: St. Martin's Press, 1993): 43–3.

30. Wilmer, "Narratives of Resistance."

31. Deloria, "If You Think about It," 68.

32. An example is the system of counting coups among the Plains nations, as well as the idea that one must pray from heart to heart between the killer and target as life is taken, whether it be the life of a human enemy or an animal for food. At the extreme is ritualized torture, but we should not necessarily equate this with the gratuitous brutality that characterizes much of the history of the modern state polity.

33. Public lecture, Carson Walks-Over-Ice, Montana State University, April 25, 1991.

34. Wilmer, *The Indigenous Voice*.

35. Deloria, "If You Think about It," 65.

36. Ibid., 64.

37. Notes, "Basic Call to Consciousness," 71.

38. Mohawk, "What We Can Learn from Indigenous Cultures."

39. Deloria, "If You Think about It," 67.

40. Mohawk, "What We Can Learn from Indigenous Cultures."

41. Issac D. Balbus, *Marxism and Domination: A Neo-Hegelian, Feminist, Psychoanalytic Theory of Sexual, Political, and Technological Domination* (Princeton, N.J.: Princeton University Press, 1982).

42. Diamond, *Fertile Ground*; and Mary Evelyn Tucker, ed., *World Views and Ecology: Religion, Philosophy, and the Environment* (Mary Knoll, N.Y.: John A. Grim, 1994).

43. Melanie Klein and Joan Riviere, *Love, Hate, and Reparation* (New York: W. W. Norton, 1964); Jacques Lacan, *The Four Fundamental Concepts of Psychoanalysis* (London: Hogarth Press, 1977); and C. Fred Alford, *The Self in Social Theory* (New Haven: Yale University Press, 1991).

44. Alford, *The Self in Social Theory*, 36.

45. Scott Coltrane, "Father-Child Relationships and the Status of Women: A Cross-Cultural Study," *American Journal of Sociology* 93, no. 5 (1988): 1060–95.

46. Ibid.

47. Ibid.; Paula Gunn Allen, *The Sacred Hoop: Recovering the Feminine in American Indian Traditions* (Boston: Beacon Press, 1986).

48. Allen, *The Sacred Hoop*; Walter L. Williams, *The Spirit and the Flesh: Sexual Diversity in American Indian Cultures* (Boston: Beacon Press, 1986); and Will Roscoe, *The Zuni Man-Woman* (Albuquerque: University of New Mexico Press, 1991).

49. Dorothy Dinnerstein, *The Mermaid and the Minotaur: Sexual Arrangements and Human Malaise* (NY: Harper and Row, 1976); and Balbus, *Marxism and Domination*.

50. Di Stefano, *Configurations of Masculinity*.

4

Sovereignty and Ecosystem Management: Clash of Concepts and Boundaries?

Veronica Ward

Introduction

Sovereignty implies control of an identifiable geographical space by the state, as the supreme legal and political authority over a physical environment and its inhabitants. Widely likened to private property rights, with all that this entails in terms of exclusive use, disposition, and control, sovereignty claims present "a conception of a world divided into separate, independent communities, delineated clearly in time and space, governed by their own sovereign authority and system of law."[1] This social world confronts a natural world "which recognizes no human sovereignties."[2] This sets the stage for a potential clash between the social and political practices that define modern state sovereignty, where differences between the "inside" and the "outside" are presumed to be clear, and the transnational application of ecosystem management approaches, where boundaries follow a very different logic and thus suggest a very different configuration of the world. As John Agnew and Stuart Corbridge note, "Space is viewed as a series of blocks defined by state territorial boundaries. Other geographical scales (local, global, etc.) are largely disregarded."[3] With the ecosystem approach the focus is on processes operating at different scales below and above the territorial state. In fact, the state, as a spatial unit, is irrelevant.

With the appearance of environmental issues on the global agenda, questions of ecologically appropriate and politically feasible management methods are the topic of extensive discussion and negotiation. These discussions raise doubts about the ability of the world of sovereign states to effectively confront and manage a natural world comprised of linked ecosystems. Whether in the realm of territorial sovereignty or in the management of open-access or common property,[4] the establishment and/or maintenance of use rights, control, and management entail authority claims. State claims to absolute authority over a given space, in conjunction with efforts to prevent closure and exclusion from "open spaces," or common-pool resources, affect efforts to conserve or preserve natural systems.[5]

The question of concern here is whether it is possible for there to be implementation of a specific type of environmental management — ecosystem management — in a world of sovereign territorial entities. "Ecosystem management," writes Edward Grumbine in his review of literature in this field, "integrates scientific knowledge of ecological relationships within a . . . sociopolitical and values framework toward the general goal of protecting native ecosystem integrity over the long term."[6] As discussions of environmental management are increasingly framed in terms of ecosystems' survival rather than in the more traditional terms of resource conservation, an analysis of the extent to which the existence of the modern sovereign state complicates the application of ecosystem principles becomes particularly timely. As the relatively new ecological approach to resource management is introduced, the question of its political viability in a world of sovereign states is well worth exploring.

I begin with a discussion of the relationship between sovereignty and territoriality, as this relationship may present a major barrier to the application of ecosystem principles, which is discussed in the next section. That section will be followed by a discussion of the two points of tension between sovereignty and ecosystem management: boundary delineation and value conflict. Analysis of these points of conflict will include a brief discussion of the Law of the Sea Convention, chosen because it demonstrates how the complexity of boundary considerations and a clash of values may work against establishment of an ecosystem regime. Despite these difficulties, there are a few cases in which ecosystem principles have been included as part of an international environmental agreement.[7] The only agreement in which efforts are proceeding, albeit slowly, toward implementation is the Convention on the Conservation of Antarctic Marine Living Resources (CCAMLR). Thus, this case provides us with an opportunity to examine the circumstances under which such an application is possible in the international arena.

Of course, any lessons drawn from the Antarctic must be treated with great circumspection as this is but one case. Even this one agreement, however, does provide some insights into those conditions that may permit or enable the implementation of ecosystem principles among states. The conclusion assesses, in light of the Antarctic experience, the potential for ecosystem management in the interstate arena.

Ecosystem management challenges the principal dimensions of state sovereignty: autonomy, control, and authority. Application of a management system on the basis of ecological principles and boundaries may require a sharing of authority, and a loss of legal and political control over flora and fauna previously considered either an integral part of the territorial state or available for the taking. The very concept of ownership that underlies claims of authority and control may be reconfigured with the establishment of ecosystem regimes. By extension, the image of the territorial state as an autonomous entity is, at the very least, thrown into question if the declared goal is one of protecting native ecosystem integrity over the long term. The integrity of the ecosystem, not the integrity of the autonomous sovereign state, is given pride of place. Not surprisingly, scientists, as chief proponents of ecosystem management, face a difficult task in persuading state officials who are resistant to any policy that threatens their states control and/or authority over valued resources. Even as the terminology of ecosystem management, like sustainable development, finds its way into official government directives and programs, its implementation will require a level of political commitment and will that may be absent. As in the Antarctic case, adoption of this approach may depend upon a confluence of interests between policymakers and scientists, a situation absent in the Law of the Sea case.

Sovereignty and Territory

Several recent studies on state sovereignty focus on conceptual "unbundling."[8] *Unbundling* refers to the decomposition of the institution of *sovereignty* and its physical representation, *territoriality*. Even as sovereignty is challenged, states may retain, or attempt to retain, degrees of control and authority over their territories. *Authority* is "the claim to exclusive right to make rules," while *control* is the capability of enforcing that claim."[9] Unbundling permits us to explore the status and content of sovereign rights. This, in turn, allows us to trace the evolution of sovereignty and its possible impact on ecological management in the interstices among and beyond sovereign spaces.

Friedrich Kratochwil, in his comparison of state sovereignty to private property, suggests that we may be witnessing an unbundling of sovereign rights comparable to that of corporate ownership in the domestic arena. "In corporations the right to control is divorced from the right to manage and from the right to receive an income from the activities of the firm." As evidence of this fragmentation, he cites the process of European integration, understood as the unfolding of "a novel notion of authority," as well as the assignment of rights in the Exclusive Economic Zones (EEZs) codified in the United Nations Convention on the Law of the Sea (LOS).[10]

In addition to *nonterritorial functional space*, defined in part by commercial economic transactions, writes John Ruggie, there are *open-access spaces* possessed by no one, such as the oceans and atmosphere, or common property resources such as transboundary waterways.[11] For Ruggie, the question is whether the modern system of sovereign states is yielding to different configurations of political space. The thrust of his argument is to suggest the emergence of a nonexclusive form of territoriality, as was the case with the medieval system of rule. Ruggie remarks particularly on the "transformative potential of global ecology" with its potential to comprise a new and very different social *episteme* — a new set of spatial, metaphysical, and doctrinal constructs through which the visualization of collective existence on the planet is shaped. This episteme would differ in form from modern territoriality and its accoutrements insofar as the underlying structural premise of ecology is holism and mutual dependence on parts.

Identifying such a transformation is difficult, as Ruggie notes. He cites such evidence as the emergence of alternative principles such as "international custodianship," which would require the state to act "in a manner that expresses not merely its own interests and preferences but also its role as the embodiment and enforcer of community norms."[12] Along these lines, a willingness to enter into arrangements based upon ecosystem management, detailed below, might also indicate the evolution of a nonexclusive territoriality. Nonexclusive territoriality is a

sharing or parceling of authority across preexisting "territorial forma-
tions."[13] Thus, ecosystem management, applicable to designated *ecological* entities, which themselves may cross existing state boundaries,
may well require that affected parties share the right to establish rules
of conduct to govern behavior within the ecosystem.

Sovereignty, in the form of territoriality, may, but need not, be an
inevitable barrier to the creation of "institutional means through which
the collectivity of sovereigns [seek] to compensate for the 'social defects'
that inhere in the modern construct of territoriality."[14] A vast body of
international law and functional arrangements attests to the fact that
cooperative management is not impossibile, although its meaning is
open to interpretation. For some, states' membership in international
organizations and regimes entails a shift in the location of authority, and
therefore undercuts their sovereignty. For others, sovereignty remains
undiminished because states freely join such cooperative arrangements
and ultimate authority remains in the hands of each state. For our
purposes, the relevant issue is the extent to which state sovereignty
presents a barrier to the establishment and effective implementation of
an ecosystem management approach.

As noted earlier, the immediate challenge is a possible reconfiguration
of space with the integrity of the ecosystem assigned precedence over
existing political borders in terms of policy formulation and direction.
As the extensive discussion that follows will make clear, control over
designated ecosystems would be shared; thus, designation of certain
territory as the responsibility of one party or the other would become
moot. Ecological links could make identification of the "outside" from
the "inside" of the state questionable, if not irrelevant, for management
purposes. This is not meant to suggest two systems of boundaries cannot
operate, but this would present serious difficulties for state leaders long
accustomed to the principle of territorial exclusivity. Before exploring
these tensions more fully, an exposition of ecosystem management is
necessary.

Ecosystem Management

Even as international relations scholars attempt to "deconstruct"
sovereignty, conservation biologists try to "reconstruct" and preserve

ecosystems. The movement away from traditional single-species resource management toward a holistic management approach, which can be traced to the 1930s when specialists urged the creation of nature sanctuaries, has gained momentum in recent years as an increasing portion of the scientific community has come to adopt this approach. Grumbine cites several reasons for this heightened interest: the perception of biodiversity crises, the failure of earlier policy initiatives to slow ecological degradation, the emerging field of conservation biology, and shifting social values regarding the relationship between human beings and nature.[15] Ecosystem management, although sharing concern for the integrity of natural systems, should not be identified simply as a branch of environmentalism. As a distinct field of study, conservation biology represents a scientific effort to better understand and maintain, within existing social parameters, natural systems valued not only in their own right, but also those recognized as essential for the sustainability of human life. Ecosystem analysts acknowledge the necessity to accommodate human communities and values in management efforts.

Conservation biologists have taken the lead in efforts to apply ecosystem management. This relatively new field of applied science "differs from other natural-resource fields," according to Grumbine in his review of the field, "by accenting ecology over economics." Unlike traditional resources management, with its emphasis on conservation for human use and consumption, it "studies biodiversity and the dynamics of extinction." Attention is paid to "how genes, species, ecosystems, and landscapes interact, and how human activities affect changes in ecosystem components, patterns and processes."[16]

To manage an ecosystem, or utilize ecosystem principles, however, as with a sovereign state, boundaries must be known; managers and policymakers must be able to identify and agree upon the entity to be conserved. This leads to a series of questions that are addressed in this section: What is an ecosystem? What phenomena comprise an ecosystem? How can its boundaries be identified? Once ecosystems are identified the question then becomes: Are there specific management goals? If so, what are they? How are they to be achieved?

There are various definitions of *ecosystem* but all include "the idea that the ecosystem includes the physical or abiotic environment in addition to biological [biotic] components (e.g., organisms)."[17] Some biologists use the terms *community* and *ecosystem* interchangeably. Anthony King and Edward O. Wilson argue, however, that inclusion of the surrounding environment distinguishes an ecosystem from a community defined as the organisms, species, or populations that inhabit a particular location.[18] *Ecosystem* will be used in this chapter to refer to a spatially bounded environment and its biotic communities.

Ecosystems entail three basic elements: ecosystem *components*, which are the inhabiting species; ecosystem *structure*, which refers to the physical patterns of the life-forms of which there may be multiple layers; and ecosystem *functions*, which are the dynamics of matter and energy processing and transfer such as carbon or nutrient cycles, hydrological cycles, and natural disturbances (i.e., fires, floods, etc.).[19] King identifies two approaches to the study of ecosystems: (1) the population community approach, which emphasizes species populations and interactions such as competition and predation; and (2) the process-functional approach, which emphasizes the transfer and processing of matter and energy. The approach taken affects the judgments reached concerning the relative integrity of any particular ecosystem. If looked at from the population perspective, the loss of any component or a change in any interaction can be judged a loss of ecosystem integrity. Alternatively, if ecosystem integrity is identified with functional integrity, with maintenance of the system's "integrated dynamic," then the loss of a species with little subsequent change in system function would result in a different conclusion. Thus, the criteria used for observing a system will affect the answer to the question: Is there a loss of ecosystem integrity?[20]

One possible bridge between these two perspectives is "keystone" species, identified as those "which play pivotal roles in their ecosystems and upon which a large part of the community depends."[21] Krill, for example, appear to play such a role in the Antarctic. Unfortunately, the term *keystone* is poorly defined. "Instead of a dichotomy of keystones and nonkeystones, communities may be better characterized by a wide range of interactions of variable strengths."[22] This further complicates the management process.

Even with general agreement on the characteristics of an ecosystem, its spatial and temporal boundaries, or its scale, must be identified.[23] Description must precede analysis and management. Differences in the extent and scale of observation will lead to different characterizations based on different components, interactions, and dynamics. Consequently, judgments on ecosystem integrity will be linked to the boundaries of the system.[24] Boundaries can coincide with a preexisting management unit such as a park; for example, much analysis is conducted around the Yellowstone system defined by park boundaries. The danger is that important system attributes may extend beyond the existing administrative unit, which is certainly the case with large animals such as grizzlies whose effective range is well beyond park boundaries. "Restricting the system to an extent less than the minimum required for interactions to occur can impact system function and may lead to a loss of ecosystem integrity."[25] Identification of unambiguous boundaries is quite difficult. "Setting boundaries," writes Grumbine, "depends on the question being asked, the particular species present, and the natural disturbances in operation."[26]

Further complications surround the issue of *greater ecosystems*, that is, interrelated ecosystems. Many animals depend upon resources that cut across different ecosystems; identification and definition of greater ecosystems is difficult. Grumbine reports that estimates of a Greater Yellowstone ecosystem range from six million to eighteen million acres.[27] One option is to identify and use "natural scales of integration . . . by considering the interactions that bind a system together and the medium of that interaction (nutrients and water flow, etc.)."[28] Where such gradients are gradual, however, there is a certain arbitrariness in defining boundaries. Time is also an important factor; observations must extend over a long enough period of time to gain a good sense of the "normal" operation of a system. In the political realm, however, time is often quite limited.

There remains the not-inconsiderable task of devising measures of integrity that will provide indicators of degradation. Conservation biologists have identified both deterministic and uncertain (stochastic) elements in extinctions; while the former are relatively easy to predict, the latter are not. Biologists must deal with a host of uncertainties: genetic, demographic, environmental, and catastrophic. Not surprisingly, no "synergistic" model exists.[29]

In the end, King believes, whatever measures are ultimately employed, they will "almost invariably come from scales and levels of organization smaller, finer, or lower than the entire system."[30] The fact that nature is dynamic, not static, that nature is not contrary to earlier thinking, in a state of balanced harmony, means effective management must be flexible. Grumbine, quoting Agee and Johnson, argues that since "what is natural 'cannot be scientifically resolved,'" management goals must rest on achieving "'socially desirable conditions.'"[31] As the ecosystemic approach includes biological and social components, to speak of "maintaining 'natural' processes," write Agee and Johnson, is problematic.

> The word "natural" remains difficult to define because it incorporates value judgments that cannot be scientifically resolved. If natural process management is assumed to mean evolution free of human influence, implementation of natural process management . . . will be difficult to accomplish.[32]

Thus, we arrive at a point of social choice, and protection of ecosystems will be just one of many possible social choices.

Despite these difficulties, management goals have been articulated by biologists. Grumbine, in his review article, identifies five goals widely endorsed by conservation biologists: (1) maintenance of viable populations of all native species in a specified area; (2) representation of all native ecosystem types across their natural range of variation in designated protected zones; (3) maintenance of evolutionary and ecological processes; (4) management over periods of time long enough to maintain the evolutionary potential of species and ecosystems; (5) the accommodation of human use and occupancy within these constraints. Grumbine also provides the working definition of *ecosystem management* cited earlier: "Ecosystem management integrates scientific knowledge of ecological relationships within a complex sociopolitical and values framework toward the general goal of protecting native ecosystem integrity over the long term." These are ambitious objectives. As Grumbine points out, the first four goals, although derived from current scientific knowledge with the intent to slow or reverse a biodiversity crisis, are as much value statements as the fifth goal.[33] The extent to which these values conflict with the values embedded in the institution of sovereignty is the issue to which we now turn.

Sovereignty and Ecosystems

There are two areas that may create tension between the claims and practices of sovereignty and ecosystem integrity: the delineation of boundaries, and the principles and goals that define the ecosystem approach. Boundary conflict is possible when a designated ecosystem cuts across existing political borders or recognized open-access areas, while conflicts can occur between the value of the integrity of biophysical systems and the value of resources for human consumption and economic returns. The two are closely linked in the real world; however, for purposes of exposition, each will be discussed separately here. The Law of the Sea provides an illustration of the complexity of boundary considerations and the difficulties inherent in ecosystem application, even in areas unclaimed by any state.

Boundary Considerations

Territorial claims do not *negate* the possibility of cooperative endeavors. Examples abound of functional regimes which enable member states to achieve or protect some valued good. This is not to suggest that reaching agreement is easy; it is usually quite difficult, requiring that state leaders either untie the bundle of rights that comprise territorial sovereignty (comparable to private property), or that they accede to a redefinition of existing open-access rights or common-pool resources (as with common property). The two situations present state officials with different issues with respect to territorial claims. With the former, ownership and control are not usually at issue, although there is, as Philip Allott notes, a broad spectrum of possible relationships between states and all kinds of things: ownership of many different degrees, possession, constructive possession, custody, control, and so on. Each of these relationships entails a bundle of rights: right to exclude, right to possess, right to alienate, right to profit, right to bequeath.[34]

When an identifiable ecosystem cuts across the boundary between two states, then the issue is whether either party is willing to engage in joint or compatible management.[35] Here the occurrence of limits or transfers of sovereignty is a function of the nature of the management regime agreed to by the parties. International law may provide grounds for claims of damages incurred due to earlier inadequate management, e.g., the well-known Trail Smelter case, thus setting limits on rights of ownership and autonomy. The likelihood of this occurring with respect to ecosystem protection would be quite small, however.[36]

Douglas Johnston reports that international river law has developed away from the principle of *unrestricted territorial sovereignty* through *unrestricted territorial integrity*, meaning natural water flow, to the emerging principle of *limited territorial sovereignty*, where community interest and equitable apportionment guide policy.[37] In this case, ownership and control rights have been limited in order to protect valuable resources, resources that are often classified as ecosystems.

But when an ecosystem lies within a state's territory, ecosystem management is unlikely to occur through international action. The Brazilian Amazonia, the largest rain forest in the world,[38] has been treated for analytical and ecological purposes as a case of an ecosystem within the boundaries of a single state.[39] Despite the urgings of outside parties, the sovereign rights of the Brazilian government appear intact. Unless voluntarily ceded, for example, through debt-for-nature swaps, something Brazilian officials have to date resisted, territorial sovereignty blocks the application of ecosystem management through international action.[40]

In the United States, efforts to devise and implement ecosystem management are on the rise through the establishment of ecosystem areas under both government and private auspices. Since these areas have only recently been created, no judgment of their relative success is yet possible. The possibility of additional designated areas is dependent upon the acquisition of funding and essential data from the federal government.[41] In addition, as with efforts at the interstate level, multiple jurisdictions (federal, state, county, local) have complicated and stalled some efforts. Several researchers note the further problems caused by the legal intervention of government agencies shaped by different missions, histories, management traditions, and rivalries.[42] One discussion of ecosystem management possibilities in Yellowstone concludes, "There is no common approach to land management. Moreover, there is no single entity empowered to assess the larger ecological ramifications of . . . activities within the ecosystem."[43] Thus, within a federal system such as that in the United States, with its divided authority and shared powers, difficulties found at the interstate level replicate themselves; like nested Russian dolls, each jurisdictional level opens up new sovereignty issues.[44]

Under the current norms of sovereignty with the emphasis on autonomy, authority, and control, ecosystem practices may be implemented only with the approval of state officials acting directly at the domestic level or indirectly through international agreements. This situation promises to continue as long as political identity and community are synonymous with the sovereign state.

An example using private property ownership should clarify the consequences of this situation. *Private property* is defined as "the authority to undertake particular actions related to a specific domain."[45] For "owners," property rights include management, exclusion, and alienation. *Management* is "the right to regulate internal use patterns and transform the resource." *Exclusion* is the "right to determine who will have an access right, and how that right may be transferred." "The right to sell or lease either or both of the above" is a right of *alienation*.[46] Imagine two property owners, X and Y, with adjoining land. X approaches Y with a scheme to manage a twenty-acre piece of land — ten of which X owns and ten of which Y owns — an offer which Y is free to accept or reject. Acceptance leads to a jointly managed program based upon ecosystem principles. Jurisdictional boundaries are adjusted to permit a sharing of authority over a specific physical space for the achievement of mutual goals. A right of withdrawal for specific reasons is retained by both parties. Therefore, with rights to manage and exclude, but not alienate, X and Y are "proprietors," but not owners, of the designated ecosystem area.[47] A refusal by Y leaves X with few options. X can try persuasion, seek to change Y's incentive structure, or offer to purchase the property; however, legally, Y retains the right to refuse participation unless property rights are redefined by the relevant government institutions. A similar situation exists in interstate negotiations, although in this arena it is states that must redefine existing rights, an important boundary condition.

The situation with respect to open-access, or common property, space presents a somewhat different problem. By definition, open-access space is neither owned nor controlled by any state. The principle of spatial exclusion upon which state sovereignty rests is absent. All that is guaranteed is a right to use and/or a right to income, which may well be lost to any other individual who similarly claims a right to access. Neither do users have the power of transmissibility, that is the power to transfer the object or resources concerned. Finally, no individual user can establish the conditions under which the property is to be used and by whom; in such circumstances individuals are "authorized users" only.[48] It is this absence of rights, as much as the right of open access granted to all comers, that has led to the problems commonly identified as endemic to common property systems — overexploitation and degradation.

These problems have been exacerbated with the advent of new technologies. No longer could they be treated as having exclusively external effects. Existing boundaries, with their respective bundles of rights, seemed inadequate as effective barriers against degradation of the oceans' resources; thus, the movement for extension of coastal state jurisdiction. This enclosure movement played out in negotiations that led to the 1982 United Nations Convention on the Law of the Sea (LOS) which entered into force in 1994. With the LOS we have a case in which bundles of rights, identified by Karen Litfin in the introduction as "those under a state's jurisdiction, those under other states' jurisdiction, and those under common jurisdiction, [are] being modified by international environmental exigencies." At the same time, the very complexity of boundary questions, along with conflicting values of sovereign rights and ecosystem principles, served to block serious consideration of an ecosystem approach. Following a discussion of the boundary issues addressed in the LOS negotiations, this chapter will explore ecosystem values and their impact on state sovereignty practices.

Law of the Sea Convention: Boundary Issues
LOS's basic objective is to establish "a legal order for the seas and oceans which will facilitate international communication, and will promote the peaceful uses of the seas and oceans, the equitable and efficient utilization of their resources, the conservation of their living resources, and the study, protection, and preservation of the marine environment."[49]

With respect to the issues of environmental quality and resource viability, the task was to deal with a public goods problem created by the externality effect of an open-access system. To prevent this public *bad*, or a "tragedy of the commons," states, through negotiations, sought to provide a public good identified as conservation of resources and marine habitat. This was to be accomplished by reducing the nonexclusive nature of the world's oceans, which entailed a reconsideration, reworking, and redefinition of sovereign rights. Unlike situations in which states are asked to move away from unrestricted territorial sovereignty, here the issue was the willingness of states to agree to the closure of previously open, free-access space. Thus, the question for state officials was whether to move toward claimant or proprietary rights. For our purposes, another question is the extent to which the assignment of rights is compatible with ecosystem management.

Rights codified under LOS cover the territorial sea, continental shelf, economic exclusion zones (EEZs), and high seas. Allott, in his legal analysis of the convention, identifies rights as *"shared powers*, shared between the holder of the power and the community of states."[50] He identifies fifty-seven kinds of legal persons and fifty-eight legal sea areas in the LOS reflected in a "layering of legal relations";[51] thus the exercise of rights by states is qualified. Absolute sovereignty, in the sense of total control, authority, and autonomy of action, is not granted, for states are authorized to act subject to limits. These limits are captured in phrases such as "must take account of," "act on the basis of," "be in conformity with," "without prejudice to."[52] For example, territorial sea sovereignty is to be exercised "subject to this Convention and to other rules of international law."[53] Multiple provisions, notes Allott, confer rights and duties on different parties that vary from one location to another.[54] Hence, the statuses of parties are judged "intrinsically interdependent." Article 58, for example, gives all states a certain status in the EEZ, described as a *horizontally shared zone*. Coastal states are not assigned exclusive jurisdiction within the two hundred-mile zone. With respect to the seabed and minerals, LOS set limits on sovereign claims, making the high seas a horizontally shared zone.[55]

How this body of complex rules, rights, and duties will be interpreted and applied remains to be seen. What is clear is that signatories were unwilling to agree to an absolute extension of sovereignty; instead they developed what has been termed a *decentralized enclosure movement*.[56] Such a situation opens the door for the application of ecosystem management.

The LOS does not, however, establish any type of ecosystem approach. Habitat preservation, particularly of critical habitats essential to preservation of all species and the maintenance of exploited species, is not mentioned in the sections of the convention that deal directly with fisheries conservation.[57] The vast majority of all ocean species are concentrated along coastlines or in shallow waters on the continental shelf. Responsibility for management in these areas was granted to coastal states, reflecting the extension of functional sovereignty.[58] In fact, many governments claimed that an extension of coastal state control was needed to assure more effective management and conservation of resources than had been achieved under regional fishery commissions.

Whether better management will follow remains an open question. What is certain is that the EEZs assure a transfer of wealth and income to nationals away from foreign operatives. Highly migratory species such as tuna, and large sharks, are not governed by specific jurisdictional rules, but under Article 64 are to be subject to global management measures both within and beyond the limits of the EEZs.[59] Such systems have yet to be established. Because of the absence of any management system, Cyrille de Klemm concludes that coastal states could destroy habitats. Nothing in the treaty, writes Klemm, "would prevent a state from draining, polluting, filling, dredging, or otherwise destroying marshes, lagoons, or estuaries which reproduce the fish."[60]

There are, however, provisions regarding habitat in the section (Part XII) of the convention on marine environment and pollution, although no explanation or definition of what is meant by the protection or preservation of the marine environment is provided. Article 194.5 does lay the legal groundwork for an ecosystem approach. It provides for a legal basis upon which coastal zone management, the preservation of unique ecosystems, and the protection of the habitats of endangered species can be developed in the future and protected areas established in all jurisdictional zones, including the high seas.[61] For this reason, Klemm judges this provision as "one of the major achievements of UNCLOS III from the point of view of conservation."[62]

In LOS we have a situation in which ownership, proprietorship, and claimant rights are legally shared through a set of complex arrangements. We also have the modification of existing bundles of rights, yet ecosystem management is largely absent. On the high seas, in particular, there are no institutional mechanisms for ecosystem management. What

explains this outcome? Certainly it was not the presence of state
sovereignty claims alone. I would argue that one key factor was the
complex nature of negotiations which led to a redefinition of rights and
reconfiguration of geographical spaces, making it highly unlikely that
state representatives would agree to the drawing of yet another set of
boundaries that would, in theory, supercede those just established.
Concerned with assuring themselves the maximum access possible in a
renegotiated space, state officials would be unwilling to consider a
management system that might further reduce their freedom of movement, control, and authority over valued resources. Here we arrive at
the second factor — conflicting values. Access to and use of valued
resources has been a defining practice of state authorities determined to
protect and enlarge state capabilities. Yet this very practice may be
circumscribed by the application of ecosystem principles.

Ecosystem Principles

As noted earlier, the ecosystem approach requires a major shift in
management techniques and philosophy. Traditional resources management (forestry, range, game, and fisheries management) is guided by the
goal of conservation to ensure the future availability of natural resources
for human beings. What maximizes production of goods and services,
whether board feet, recreation days, or fish catches, is most desirable;
hence the concern with the achievement of optimum yield or maximum
sustainable yield. The very term *resource* implies that value of a species
comes from its use; what cannot be used is judged worthless.[63]

Closely associated with natural resources management has been
single-species conservation and protection. For conservation biologists
such an approach has "proved of limited value" because of its focus on
"what are called 'charismatic megavertebrate,'" such as elk, bear, or
elephant; its piecemeal approach; its focus on threats from hunting and
trapping rather than habitat degradation; and its reactive rather than
proactive stance.[64]

Ecosystem management rests on a different set of values and beliefs
about the relationship between humans and the natural world. Conservation for human consumption or economic returns is not the highest
value. Instead, as noted, maintenance of natural ecosystems over the
longterm is the principal goal. Human needs and wants are not ignored

by the ecosystem perspective, but they are to be met in a manner consistent with the continued integrity of natural systems.[65] With the exception of the convention on living resources in the Antarctic, discussed below, there is little evidence that the values of ecosystem management have had any substantive, as opposed to rhetorical, role to play in interstate negotiations on the environment.[66] In fact, the record of LOS negotiations demonstrates the continued importance of the traditional view of resources; concern over the economic and consumptive returns to citizens and corporations dominates the negotiating positions of policymakers in most countries. The absence of marine ecologists at the LOS negotiations is telling.[67] In this instance there was no convergence of interests between scientists and political leaders.

As recently as the late 1980s, ecosystemic priorities appeared to be low on the agenda at a conference of professional managers on U.S.-Canada transboundary issues. As a participant, Grumbine reports:

No one...grounded their management in the nature of natural ecosystems. They based their decisions on socially constructed images of "natural resources" mediated by administrative politics within large public and private bureaucracies. Knowledge of viable populations, habitat fragmentation, buffer zones, and natural disturbances was not widespread here. Politics, not ecology, was our common currency.[68]

As long as traditional perspectives remain dominant among officials within those countries whose biologists have taken the lead in developing the ecosystem approach, the values articulated by conservation biologists are unlikely to be implemented at the international level.[69] Finally, the existence of gaps and uncertainties in the knowledge of conservation biologists further undercuts their influence in shaping public policy.[70] What then is it about the situation in the Antarctic that led to the adoption of an ecosystem approach?

Living Resources in the Antarctic

The area south of sixty degrees south latitude has been under the management of "Consultative Parties" to the Antarctic Treaty System (ATS) since 1959. Management authority lies in the hands of the twenty-six Consultative Parties; an additional fifteen non-Consultative Party members do not have full voting rights.[71] Intervening decades have witnessed the negotiation of numerous agreements and understandings which cover a host of substantive and procedural matters; many deal with the preservation and/or conservation of the Antarctic environment and its flora and fauna.[72]

The Convention on the Conservation of Antarctic Marine Living Resources (CCAMLR), which entered into force in 1982, five years after negotiations first began, is one of the few international resource regimes to incorporate an ecosystemic approach. It is, as Klemm writes, the only fisheries treaty that functions as an ecological treaty as well.[73] A reading of the relevant article (II) makes clear, as J. A. Gulland notes, that a compromise was reached between "two groups of interest," one concerned with conservation, or rational exploitation, and the other with preservation.[74] Article II calls for "maintenance of ecological relationships between harvested, dependent, and related populations" and the "prevention of changes or minimization of risk of changes in the marine ecosystems which are not potentially reversible over two or three decades." Thus, the CCAMLR appears to represent an ecosystemic approach, while simultaneously allowing for the extraction of resources. Two questions arise: What explains this outcome? Has it worked? To answer the first, we begin with a description of the status of sovereignty claims in the Antarctic.

Antarctica and its surroundings are subject to conflicting territorial claims; sovereignty has been and remains a politically sensitive issue, even more so with the advent of resource issues. Seven of the twenty-six Consultative Parties lay claim to sectors of the continent with three — Argentina, Chile, and Great Britain — making overlapping claims. The remaining members either refuse to recognize these claims or retain rights to future sovereignty claims.[75] Article IV of the treaty has effectively "frozen" claims in an effort to protect the positions of claimant and nonclaimant states alike.[76] Thus, in the Antarctic we have a situation that seems to encompass some elements of private property, open-access, and common property systems. The very ambiguity of territorial claims encouraged a movement toward the establishment of a resource regime.

With the advent of new fishing technology and the emergence of distant water fishing fleets by two Consultative Party members, the former Soviet Union and Japan, concern arose over the impact of commercial activity on the status of other members' territorial claims. (Recall that somewhat similar concerns led to LOS negotiations.) These two states appeared to be on the brink of initiating full-scale commercial harvesting of Antarctic krill (Antarctic shrimp). Furthermore, there was the possibility that claimant states might declare two hundred-mile EEZs, thereby effectively destroying the ATS. Those who asserted sovereignty wished to protect their rights, including coastal state jurisdiction, while those who denied such assertions wished to prevent their exclusion from access. The issue was resolved through the "bifocal" language of Article IV(2)b of the CCAMLR, which stipulates that nothing in the convention shall be interpreted as prejudicing claims to coastal state jurisdiction.

Different interpretations of this section are possible because there is no sovereignty dispute over the territory that is within the convention area but north of sixty degrees south. The convention covers the area up to the Antarctic convergence which, at certain points, extends beyond sixty degrees south latitude, the area that comprises the Antarctic Treaty System. But, of course, the territory south of sixty degrees south is subject to dispute. Therefore, claimant states interpret this article as protection for their coastal state jurisdiction, while nonclaimants interpret this article as applicable only to the undisputed territory north of sixty degrees south. James Barnes, a member of the U.S. negotiating team, states that many, if not all, members saw the negotiating process as a way to solidify their territorial claims.[77]

Even as member states moved to protect their respective territorial positions, they were also concerned about environmental integrity as Japan and the Soviet Union increased their fishing activities. The ramifications of overharvesting of krill are not fully known, but it is known that krill, located at the lowest level of a complicated food chain, are an essential element in the Antarctic ecosystem. As a "keystone species," krill are the main food source for nearly all larger marine species, including whales. There was general acceptance of the role of krill and the possible consequences of overexploitation, even from harvesting countries like the Soviet Union. The United States was the initiator and leading proponent of the ecosystem approach, with strong support coming from Great Britain, Norway, and New Zealand. The strongest resistance, not surprisingly, came from Japan and the Soviet

Union. In the end, both harvesting states went along with the inclusion of ecosystem management in the convention. Mahinda Perera reports, based on her reading of available documents, that the ecosystem was "unworkable" given the current state of knowledge, although it played a role in fishing states' acceptance of the agreement.[78] In addition, reference to "rational use" as part of the definition of *conservation* in the treaty served to provide a degree of protection against harvesting prohibitions. These circumstances led to "conditional acceptance" of the ecosystem standard by the Soviet and Japanese delegations.[79] The U.S. delegation was able to push this innovative approach because its fishing industry had shown little interest in harvesting Antarctic marine resources. At the same time, the decision-making system incorporated into the convention, based on consensus, lessened any fears that Soviet or Japanese officials may have had about the possible establishment of unacceptable conservation standards.

To further quiet any sovereignty-based concerns, enforcement responsibility under CCAMLR was placed in the hands of member states (Article XXI). In light of sovereignty ambiguities, the commission created to oversee CCAMLR was granted no inspection or enforcement powers, an emission suggesting that compliance may be less than total.[80] Worries over ineffective or incomplete compliance are therefore not misplaced.[81] Thus, it was the specific configuration of interests in the Antarctic, along with an acknowledgment of potentially negative environmental and economic consequences, which resulted in the establishment of this ecosystem management regime. Here scientists found an important political ally in the United States at a time when there appeared to be few costs associated with adoption of this approach. Unlike the LOS case, in the Antarctic there were no changes in existing "understandings" of sovereignty claims. Treaty members chose to maintain their long-standing position on boundaries captured in both the "bifocal" clause and Article IV of the original 1959 treaty.[82]

Has CCAMLR worked? The answer seems to be that it is too soon to tell, but the indications are promising, and, for some observers, there is reason for optimism. Agreement on ecosystem management was made easier given a consensus on boundaries. The ecosystem encompasses the Antarctic convergence where "cold polar waters submerge beneath the warmer edges of the Atlantic, Pacific, and Indian oceans." The convergence, which marks two biologically distinct bodies of water, acts as a barrier to the movement of species and provides an identifiable boundary for management purposes.[83]

Early concerns that consensus-based decision making would prove a political barrier to effective action initially appeared correct as advocates of conservation measures found their way blocked during the first few years of operation. Francisco Orrego Vicuna reports, however, that from 1987 on, a number of conservation measures have been adopted, and the consensus rule has ceased to be used for political purposes.[84] Perera dates the "enhanced effectiveness of CCAMLR" from 1990, as reflected in the adoption of "a record number of conservation measures." John Heap, United Kingdom commissioner to convention meetings, concludes that "CCAMLR is working and . . . its effectiveness is improving."[85] One of the reasons cited by Perera for this improvement is the implementation of enforcement mechanisms.[86]

Of forty-three measures adopted during the first decade of CCAMLR, only seventeen were instituted in the first eight years.[87] During the first phase (1982–89) fishing states used scientific uncertainties to resist or oppose regulatory measures while trying to base policy on a species-by-species, rather than an ecosystemic, approach.[88] In 1990, however, progress began with the implementation of enforcement mechanisms, along with improvements in procedures for the acquisition of data and additional sources of expertise.[89]

Some nonfishing states have increased their scientific research and logistical capabilities as a means to play a more assertive role in the regime. Enhanced scientific understandings of the ecosystem's dynamics, though still far from complete, have offset the position of fishing states that "rational use" could justify continued operations. Establishment of "protected sites for monitoring the ecosystem" demonstrates the increased bargaining power of nonfishing states.[90] At the same time, post-1989 changes in Eastern Europe, and the dissolution of the Soviet Union, proved beneficial for conservation efforts. Finally, difficulties associated with processing and marketing krill reduced their economic appeal, which in turn led to a decline in the size of catches.[91]

Effective ecosystem management remains elusive, however, even as political commitment is on the rise, because knowledge of species and their interactions is incomplete. CCAMLR is operating on the frontiers of science. The complexities associated with efforts to conserve and protect multiple, interdependent species can be overwhelming.[92] Vladimir Korzun notes: "The convention did not specify the sorts of data fishers should report, for the good reason that no one was quite sure what would be needed."[93] Once data are available, the next problem is determining what conclusions should be drawn. In particular, knowledge of krill, the primary focus of concern for the convention, remains limited. Essential facts such as their lifespan and population sizes remain contested, making it difficult to formulate limits for krill taking. Finally, along with the many scientific uncertainties, there remain serious weaknesses in the inspection and observation system.[94]

Despite these difficulties and limitations, the general feeling seems to be one of optimism; efforts are being made to establish an effective ecosystem management system, which continues to set the ATS apart from other international agreements. It is too soon to judge whether success in terms of ecosystem integrity will be achieved.

Conclusion

What insights can we draw from the Antarctic case regarding conditions that may permit the implementation of ecosystem principles in a world of sovereign states? The first and most obvious point is the relative ease of identification of acceptable ecosystem boundaries, which certainly smoothed the way to agreement. While there may be other ecosystems that could be easily demarcated, it would seem more likely that most potential ecosystem boundaries would be unclear and subject to dispute, given scientific uncertainties. (Recall, for example, differences over a Greater Yellowstone ecosystem.) Thus the Antarctic, with its clearly delineated convergence, may be, if not unique, relatively uncommon.

The practices and associated goals of state sovereignty are a second boundary consideration. Even where ecosystems are readily apparent, the "demands" of state sovereignty may prove decisive. In the Antarctic there is a tangled web of sovereignty claims and denial of claims. Neither claimants nor nonclaimants undertook to force a change in the prevailing situation, even in order to strengthen their own respective territorial positions. Unlike the two hundred-mile EEZ claims, claimants were willing to settle for the "bifocal" language. This made easier acceptance of a management system that requires a "holistic" approach.[95] The so-called freezing of claims, which has facilitated a series of agreements among Antarctic Treaty System parties, held. As I have argued elsewhere, member states are committed to the continuance of the existing treaty system, which grants authority and control to a relatively small number of states.[96] This served to mitigate any serious challenges on sovereignty grounds. Here again, replication of such a set of circumstances is not impossible, but the likelihood is slight.

Ecosystem management could be applied in LOS, although the prospects are not hopeful. Recall that Ruggie hypothesized that global ecology may provide support for nonexclusive territoriality through a commitment to international custodianship or stewardship. Ecosystem management would certainly reflect such a commitment. LOS, with its various types and degrees of authority and its treatment of the seabed as the "common heritage of mankind," does represent nonexclusivity and at least the rhetoric of stewardship. But LOS does not entail an ecosystem management program, for the reasons described above.

Thus, to answer the query posed in the title of this chapter, sovereignty and ecosystem management do involve a clash of concepts and boundaries. These conceptual and physical conflicts are, not surprisingly, closely intertwined. The difference in the spatial configurations of territorial sovereignty and ecosystems is reflected in the differences in value priorities. The widespread goal of governments to assure the greatest economic returns for themselves and/or their citizens in the name of continued state survival and "growth" is an objective that does not even appear among conservation biologists' management goals. The project of turning nature into exploitable resources, as chapter 2 in this volume demonstrates, has been a central concern for the modern state. Territorial claims can make negotiations complex and difficult as each state seeks to maximize its authority over policy decisions and valued goods. As long as the values of traditional resources management—maximization of economic returns and human consumption—prevail, the scope of ecosystem management will remain severely limited. It was because such values were not of central concern in the Antarctic, particularly for certain key member states such as the United States, that it was possible to create this agreement based on ecosystem management. In the immediate future, as efforts to implement the Convention on Biodiversity proceed, we may witness repeated challenges to state sovereignty in the name of ecosystem integrity. An ecological approach to protecting biological diversity would certainly complicate an already complex set of negotiations which encompass much of the physical world. However, without a reconfiguration of priorities, itself tied to a spatial reconfiguration, the future of ecosystem management is bleak. The "new and very different social episteme" that Ruggie associates with global ecology remains at the margins of interstate negotiations.

Notes

1. Joseph A. Camilleri and Jim Falk, *The End of Sovereignty? The Politics of a Shrinking and Fragmenting World* (Aldershot, England: Edward Elgar Publishing, 1992): 172.

2. Patricia M. Mische, "Ecological Security and the Need to Reconceptualize Sovereignty," *Alternatives* 14 (1989): 389.

3. John Agnew and Stuart Corbridge, *Mastering Space: Hegemony, Territory, and International Political Economy* (New York: Routledge, 1995): 79.

4. *Open access* refers to situations in which nobody is excluded from use. *Common property* refers to situations of common ownership or collective control.

5. Common-pool resources are resources "used by multiple individuals regardless of the type of property rights involved." Edella Schlager and Elinor Ostrom, "Property Rights and Natural Resources: A Conceptual Analysis," *Land Economics* 68 (1992), 249.

6. R. Edward Grumbine, "What Is Ecosystem Management?" *Conservation Biology* 8 (1994): 31.

7. Cyrille de Klemm reports that many early twentieth-century treaties which aimed to conserve specific flora and fauna often referred to "their natural habitats," as well as making reference to "ecosystems." Today, the Convention on Wetlands of International Importance Especially as Waterfowl Habitat (1971), the Convention Concerning the Protection of the World Cultural and Natural Heritage (1972), and the Great Lakes Water Quality Agreement (1978) all include reference to conservation of ecosystems. However, in none of these agreements has an effort been made to implement an ecosystem management scheme. Cyrille de Klemm, in collaboration with Clare Shine, *Biological Diversity Conservation and the Law: Legal Mechanisms for Conserving Species and Ecosystems* (Cambridge, England: IUCN, 1993): 7–14.

8. Friedrich Kratochwil, "Of Systems, Boundaries, and Territoriality: An Inquiry into the Formation of the State System," *World Politics* 39, no. 1 (1986): 27–52. Kratochwil, "The Concept of Sovereignty: Sovereignty as

Property," unpublished paper, May 1992, later published as "Sovereignty as Dominium: Is There a Right of Humanitarian Intervention?" in Gene M. Lyons and Michael Mastanduno, eds, *Beyond Westphalia? State Sovereignty and International Intervention* (Baltimore: Johns Hopkins University Press, 1995); John Gerard Ruggie, "Territoriality and Beyond: Problematizing Modernity in International Relations," *International Organization* 47, no. 1 (1993): 139—74; Janice E. Thomson, "State Sovereignty and International Relations: Bridging the Gap between Theory and Empirical Research," *International Studies Quarterly* 39 (Summer 1995): 213–33.

9. Thomson, "State Sovereignty," 223.
10. Kratochwil, "Concept of Sovereignty," 25, 26. In the published version "Sovereignty as Dominium," 28.
11. Ruggie, "Territoriality and Beyond," 164.
12. Ibid., 173.
13. Ibid., 150.
14. Ibid., 171.
15. Ibid., 29.
16. Ibid., 29. An early and well-known exponent of conservation biology is the biologist E. O. Wilson.
17. Anthony W. King, "Considerations of Scale and Hierarchy," in Stephen Woodley, James Kay, and George Francis, eds., *Ecological Integrity and the Management of Ecosystems* (Ottawa, Canada: St. Lucie Press, 1993): 20.
18. Ibid., Edward O. Wilson, *The Diversity of Life* (New York: W.W. Norton, 1992): 396.
19. R. Edward Grumbine, *Ghost Bears: Exploring the Biodiversity Crisis* (Washington, D.C.: Island Press, 1992): 22; King, "Considerations," 21.
20. King, "Considerations," 25–27.
21. Reed F. Noss and Allen Y. Cooperrider, *Saving Nature's Legacy: Protecting and Restoring Biodiversity* (Washington, D.C.: Island Press, 1994): 7.
22. Ibid., 8.
23. King, "Considerations," 28.
24. Ibid., 28.
25. Ibid., 29.
26. Grumbine, *Ghost Bears*, 58.
27. Ibid., 58.
28. King, "Considerations," 38.
29. Ibid., 32. See Grumbine, *Ghost Bears*, 33–37, for a discussion of the uncertainties associated with each of processes.
30. King, "Considerations," 36.
31. James K. Agee and Darryll R. Johnson, quoted in Grumbine, *Ghost Bears*, 158.

32. James K. Agee and Darryll R. Johnson, "Introduction to Ecosystem Management," in Agee and Johnson, eds., *Ecosystem Management for Parks and Wilderness* (Seattle: University of Washington Press, 1988): 10.

33. Grumbine, "What Is Ecosystem Management?," 31. Grumbine is careful to note that ecosystem management is an "evolving concept" subject to different interpretations. However, after a review of the major journals in the field, he is able to report on several goals that most biologists endorse.

34. Philip Allott, "Power Sharing in the Law of the Sea," *American Journal of International Law* 77 (1983): 9.

35. This situation is similar to efforts to synchronize or harmonize macroeconomic policies among governments.

36. The *Trail Smelter* case, decided in 1941, required the Canadian government to pay damages to the U.S. government to recover the costs associated with pollution from a smeltering plant in British Columbia judged responsible for damages in the state of Washington. In the decision the arbitration panel stated: "No state has the right to use or permit the use of its territory in such a manner as to cause injury by fumes in or to the territory of another or the properties or persons therein, when the case is of serious consequence and the injury is established by clear and convincing evidence." For a general discussion of the legal status and ramifications of this decision see Patricia W. Birnie and Alan E. Boyle, *International Law and the Environment* (Oxford: Clarendon Press, 1992): chapter 3.

37. Douglas M. Johnston, "The Environmental Law of the Sea: Historical Development," in Douglas M. Johnston, ed., *The Environmental Law of the Sea* (Gland, Switzerland: International Union for Conservation of Nature and Natural Resources, 1981): 22.

38. Andrew Hurrell, "Brazil and the International Politics of Amazonian Deforestation," in Andrew Hurrell and Benedict Kingsburg, eds., *The International Politics of the Environment* (Oxford: Clarendon Press, 1992): 400.

39. The Amazon Basin actually extends into Brazil, Bolivia, Colombia, Ecuador, Peru, and Venezuela. Some observers would also include French Guiana, Guyana, and Surinam.

40. Christopher Stone notes that under current international law outsiders have little recourse in the absence of explicit bilateral or multilateral treaties. See Christopher D. Stone, *The Gnat Is Older than Man: Global Environment and Human Agenda* (Princeton, N.J.: Princeton University Press, 1993): 38.

41. Personal conversation with an official of the Wilderness Society, Washington, D.C., office, in February 1996. For a detailed study of ecosystem management in the United States see Steven L. Yaffee, et al., eds., *Ecosystem Management in the United States: An Assessment of Current Experience* (Washington, D.C.: Island Press, forthcoming).

42. See Robert B. Keiter, "Beyond the Boundary Line: Constructing a Law of EcosystemL Management," in *University of Colorado Law Review* 65 (1994): 301. T. W. Clark, E. D. Amato, D. G. Wittenmore, and A. H.

Harvey, "Policy and Programs for Ecosystem Management in the Greater Yellowstone Ecosystem: An Analysis," in *Conservation Biology* 5 (1991): 415.

43. Robert B. Keiter, "Taking Account of the Ecosystem on the Public Domain: Law and Ecology in the Greater Yellowstone Region," in *University of Colorado Law Review* 60 (1989): 985.

44. An interesting variant of this type of case is the question of the status of a state's territorial sea.

45. Schlager and Ostrom, "Property-Rights and Natural Resources," 250.

46. Ibid., 251.

47. Ibid., 253.

48. Ibid., 252.

49. Law of the Sea Convention, preamble, paragraph 4.

50. Allott, "Power Sharing," 27.

51. Ibid., 10.

52. Oscar Schachter, "Concepts and Realities in the New Law of the Sea," in Giulio Pontecorvo, ed., *The New Order of the Oceans: The Advent of a Managed Environment* (New York: Columbia University Press, 1986): 38.

53. Law of the Sea Convention, article 2.

54. Allott, "Power Sharing," 13.

55. Ibid., 16.

56. For a detailed discussion of these issues see Robert L. Friedheim, *Negotiating the New Ocean Regime* (Columbia: University of South Carolina, 1993).

57. Cyrille de Klemm, "Living Resources of the Ocean," in Johnston, *Environmental Law of the Sea*, 135.

58. Ibid., 123.

59. Klemm, "Living Resources," 72, 124.

60. Ibid., 235–36.

61. Ibid., 137.

62. Ibid., 137.

63. See Noss and Cooperrider, *Saving Nature's Legacy*, 72.

64. Ibid., 70, 71.

65. See Noss and Cooperrider, 85, for a detailed comparison of the two perspectives.

66. In 1978 an ecosystem approach to management of the Great Lakes Basin was proposed by the International Joint Commission Great Lakes Research Advisory Board. Elements of an ecosystem approach were incorporated into the Great Lakes Water Quality Agreement signed on October 22, 1978. Reportedly, conceptual support has not yet been turned into specific

programs or mechanisms to institute implementation. See Lynton K. Cald-
well, ed., *Perspectives on Ecosystem Management for the Great Lakes: A
Reader* (Albany: State University of New York, 1988); "Great Lakes Health
Rests with Management," *Water Environment and Technology* 7 (January
1995): 23; J. Vallentyne and A. Beeton, "The Ecosystem Approach to
Managing Human Uses and Abuses of Natural Resources in the Great Lakes
Basin," *Environmental Conservation* 15 (1988): 58–62.

67. See Friedheim, *Negotiating the New Ocean Regime*, and Klemm, "Living
Resources."

68. Grumbine, *Ghost Bears*, 132.

69. Kempton, Boster, and Hartley, in their recent study on *Environmental
Values in American Culture*, note the existence of a strong, deeply felt
commitment to environmental values, including "biocentric" values which
grant nature intrinsic rights. Their interviews, however, "provide no evidence of an ethic to preserve ecosystems rather than isolated species." "Our
lay [nonscientists] informants seemed to think that species preservation
required refraining from directly killing endangered animals, rather than by
protecting their habitats as refuge areas. They have not yet understood that
species are endangered not so much by the gun as by the development plan,
the chainsaw, and the bulldozer." Willett Kempton, James S. Boster, and
Jennifer A. Hartley, *Environmental Values in American Culture* (Cambridge, Mass.: MIT Press, 1995): 112.

70. The possible links between scientific knowledge and the formation of
international cooperative agreements are explored in the work of both Ernst
Haas and Peter Haas. See Ernst Haas, "Is There a Hole in the Whole?
Knowledge, Technology, Interdependence, and the Construction of Interna-
tional Regimes," *International Organization* 29 (1975): 827–76; Ernst
Haas, "Why Collaborate? Issue-Linkage and International Regimes," *World
Politics* 32 (1980): 357–405; Peter Haas, "Do Regimes Matter? Epistemic
Communities and Mediterranean Pollution Control," *International Organi-
zation* 43 (1989): 377–403; and Peter Haas, ed. special issue, "Knowledge,
Power, and International Policy Coordination," *International Organization*
46 (1992).

71. Under the conditions of the original treaty arrangements, it is only upon
fulfillment of ascension (that a state show "substantial" research activity)
that full voting rights are granted. There are, however, no set criteria for
evaluating whether an applicant state has fulfilled this requirement. Thus the
decision-making process may be open to a charge of arbitrariness.

72. The most prominent agreements are the 1964 Agreed Measures for the
Conservation of Antarctic Fauna and Flora, the 1972 Convention for the
Conservation of Antarctic Seals (CCAS), the 1980 Convention on the
Conservation of Antarctic Marine Living Resources, and the 1991 Protocol
on Environmental Protection, which has yet to enter into force.

73. Klemm, *Biological Diversity*, 49.

74. J. A. Gulland, "The Management Regime for Living Resources," in Christopher C. Joyner and Sudhir K. Chopra, eds., *The Antarctic Legal Regime* (Dordrecht, Netherlands: Martinus Nijhoff Publishers, 1988): 229.

75. Veronica Ward, "Regime Norms as 'Implicit' Third Parties: Explaining the Anglo-Argentine. Relationship," *Review of International Studies* 17 (1991): 174–75.

76. The article reads: "No acts or activities taking place while the present Treaty is in force shall constitute a basis for asserting, supporting, or denying a claim to territorial sovereignty ... or create any rights of sovereignty.... No new claim, or enlargement of an existing claim ... shall be asserted while the present Treaty is in force" (Subsection 2, Article IV).

77. James Barnes, "The Emerging Antarctic Living Resources Convention," *Proceedings of the American Society of International Law* (1979): 281.

78. Mahinda Perera, "Change and Continuity in Antarctic Environmental Protection: Politics and Policy" (Ph.D. diss., Dalhousie University, 1995): 205.

79. Ibid., 205–6.

80. Ibid.

81. Ibid.

82. Article 4 reads, in part, "No acts or activities taking place while the present treaty is in force shall constitute a basis for asserting, supporting, or denying a claim to territorial sovereignty in Antarctica or create any rights of sovereignty in Antarctica. No new claim, or enlargement of an existing claim, to territorial sovereignty in Antarctica shall be asserted while the present treaty is in force."

83. Vladimir Korzun, "The Southern Ocean," in James M. Broadus and Raphael V. Vartanov, *The Oceans and Environmental Security: Shared U.S. and Russian Perspectives* (Washington, D.C.: Island Press, 1994): 190; Arthur Watts, *International Law and the Antarctic Treaty System* (Cambridge: Grotius Publications, 1992): 151.

84. Francisco Orrego Vicuna, "The Effectiveness of the Decision-Making Machinery of CCAMLR: An Assessment," in Arnfinn Jorgensen-Dahl and Willy Ostreng, eds., *The Antarctic Treaty System in World Politics* (New York: St. Martin's Press, 1991).

85. John A. Heap, "Has CCAMLR Worked? Management Politics and Ecological Needs," in Jorgensen-Dahl and Ostreng, *Antarctic Treaty System*, 50–51; Perera, "Change and Continuity," 246–47.

86. Perera, "Change and Continuity," 246.

87. Ibid., 234.

88. Ibid., 231, 235.

89. Conservation measures established since 1990 include catch limitations, prohibitions on certain species, closure of waters, and various reporting measures. Ibid., 232–33.

90. Ibid., 248–49.

91. Ibid., 250–51, 252–53.

92. See the discussion by Marinelle Basson and John R. Beddington, "CCAMLR: The Practical Implications of an Eco-System Approach," in Jorgensen-Dahl and Ostreng, *Antarctic Treaty System.*

93. Korzun, "The Southern Ocean," 210.

94. Ibid., 212–13. In addition, see the updated discussion by Perera, "Change and Continuity," 246–49.

95. The absence of a "native" human population probably also made adoption easier.

96. Ward, "Regime Norms."

5

The Nature of Sovereignty and the Sovereignty of Nature: Problematizing the Boundaries between Self, Society, State, and System

Ronnie D. Lipschutz

Nature, to be commanded, must be obeyed.
—Sir Francis Bacon (1620)

Say what some poets will, Nature is not so much her own ever-sweet interpreter, as the mere supplier of that cunning alphabet, whereby selecting and combining as he pleases, each man reads his own peculiar lesson according to his own peculiar mind and mood.
—Herman Melville, *Pierre, or the Ambiguities*

Them what's got shall get,
Them what's not shall lose...
—Billie Holiday, "God Bless the Child"

Introduction

According to a 1995 study published by the World Bank, the "wars of the next century will be over water."[1] Such warnings are not new; by now, the invocation of "water wars" is almost commonplace, as a search of any bibliographic database will attest.[2] The bank, however, conveys special authority because of its long involvement in the planning and development of water resources. Not only does it rely on "experts" on water and its use, but as a central icon of the global economic system and a major funder of large-scale water supply systems, it must also be listened to. Yet, is the bank correct? I have written this chapter as an exegesis of the "nature of sovereignty and the sovereignty of nature," with particular reference to geopolitical discourses of "environment and conflict" in the late twentieth century.

I begin with a closer examination of the World Bank's arguments about "water wars." I then link this argument to the ideas of the classical geopolitical scholars of the nineteenth and twentieth centuries — Mackinder, Spykman, Gray — and the ways in which they sought to naturalize the relationship between geography and state power as a means of legitimizing the state's efforts to redress scarcity under anarchy through military means. I also point out how, in recent years, a parallel geopolitics of the body that puts forth various propositions about genetic scarcity and the ownership of its genetic resources has developed, in concert with the ideological shift, rooted in the hegemony of neoliberalism, from the individually sovereign state to the individually sovereign consumer. I then turn to a discussion of sovereignty and property. I argue that sovereignty is best understood as a mode of exclusion, that is, a way to draw boundaries and establish rights of property against those who would transgress against the state.

The supposed solution to this geopolitical dilemma was (and is) reliance on yet another naturalizing discourse, that of the market. Markets rely on the uneven distribution of resources; Malthus may have been right, but *scarcity*, as we define it today, is a necessary condition for markets and property rights to function. Hence, following World War II, the diffusion of "embedded liberalism" throughout the world helped to disseminate a new geopolitics that has, more recently, come to rely on the concept of economic interdependence in order to maintain the fiction, if not the fact, of political sovereignty. In recent years, as a result, we have seen the emergence of discussions of, on the one hand, "limits to growth" and "sustainability" and, on the other, "environment and security," both the direct descendants of these two geopolitical discourses. As such, these discourses assume or attempt to reinstate sovereign natural boundaries where, perhaps, none should exist.

Water Rats and Water Thieves[3]

The World Bank's arguments about the future of conflict in the Middle East leave a number of questions unanswered or at least underdeveloped. For instance, who will fight over water? The bank, and most other water scholars, believe that, "The Middle East is the likeliest crucible for future water wars. A long-term settlement between Israel and its neighbors will depend at least as much on fair allocation of water as of land. Egypt fears appropriation of the Nile's waters . . . by upstream

Sudan and Ethiopia. Iraq and Syria watch and wait as Turkey builds dams in the headwaters of the Euphrates."[4] It is clear that the combatants will be states.

Why will they fight over water? On this point, the bank's reasoning is less clear. On the one hand, water is scarce, and people, according to conventional neoliberal analysis, naturally come into conflict over scarce resources. On the other hand, geography (or, more precisely, nature) does not see fit to constrain rivers, drainages, and mountains within the confines of national boundaries. Indeed, rivers make excellent borders between countries because they are such prominent geographic features—although they do have a tendency to wander at times—and are difficult to cross. Hence, the application of a combination of geopolitical and neoclassical logics leads to the conclusion that, if resources are essential, scarce, and "in the wrong place," states that lack them will go to war with states that have them.[5] Q.E.D.

What, then, is the bank's solution? Here emerges a paradox. First, we are confronted with conflict and the possibility of war. Suddenly, we are transported from the "State of Nature" to the nature of markets.[6] In Nature, people fight and often come out losers; in markets, they bargain according to self-interest and come out winners. Thus, according to the bank's vice-president for the environment, the avoidance of water wars is to be found in "rational water management"—that is, in the transmogrification of neoliberalism from a doctrine of conflict arising from the maldistribution of state sovereignty over resources to one of exchange of scarce resources—in this instance, money and water—between sovereign consumers. How is this to be accomplished? Through the appropriate pricing of water at its "true" marginal cost—although the "true" marginal cost of water to people lacking sufficient liquidity or, for that matter, defending the nation is left undefined. This, it is thought, will lead to the assumption of water's "proper place as an economically valued and traded commodity" as well as efficient and sustainable use of the substance through technologies of conservation. As the author of an article in the *Economist* describing the report puts it, "The time is coming when water must be treated as a valuable resource, like oil, not a free one like air." Not, perhaps, an ideal parallel—especially insofar as the Persian Gulf War was more about the political impacts of oil prices than absolute supply[7]—but the point is well taken. It is probably better to truck and barter in natural resources than it is to fight for them—if those are the only choices.

The tension embedded in this discourse over water is, perhaps, obvious: the nature of sovereignty or control by states and peoples in possession of resources vs. the sovereignty of nature over human social organization and politics, a condition frequently characterized as *ecological interdependence*. I use the term *sovereignty* here to conform with Karen Litfin's definition of it as control, autonomy, and authority, offered in chapter 1. While the concept of state sovereignty is contested but, nonetheless, reasonably clear, Nature's sovereignty is less so. On the one hand, nature—the material world—imposes constraints on human activities and, in a sense, limits what can be freely and autonomously done. On the other hand, Nature—the reified construction that seeks to account for power and hierarchy—is invoked in order to naturalize the control, autonomy, and authority exercised by some human beings (often acting in the name of states over other human beings).

As we well know, national boundaries do not correspond to natural ones. Historically, the solutions offered to this problem have been either conflict or commerce. To paraphrase Franklin Roosevelt's secretary of state, Cordell Hull, "If water does not cross borders, soldiers will"—with both being "naturalized" via competing neoliberal visions of the State of Nature. The invocation of nature's sovereignty—whether Hobbesian or Smithian—thus becomes one more attempt by the powerful to veil relations of domination and subordination, and is simply intended to put to rest arguments about politics and social choice.

I do not believe "water wars" have occurred or are inevitable any more than I believe "ore wars" or "oil wars" litter human history.[8] Resource wars are really about "who is to be master," as Humpty Dumpty put it. It is always easier to explain a conflict or war as having been started over access to or supply of oil rather than over the local distribution of power and justice or the political costs of high gasoline prices. But the continuing fascination with the idea evinced by policymakers and the public is symptomatic of what John Agnew and Stuart Corbridge call "geopolitical discourses."[9] It is also symptomatic of what Anthony Giddens calls the "double hermeneutic" since such belief systems arise from real practices and have real effects on politics and policy. The result is that these discourses lead to bad politics and policy by naturalizing that which is truly political. Indeed, as I will argue, the concept of *ecological interdependence* is just one more geopolitical discourse, in that those who use it seek to naturalize both economic liberalism as the "true" condition of, and individuals or the state as the "true" actors in, the global system. Or, as the title of this chapter puts it, the assumed "sovereignty of nature" is used to naturalize certain propositions about the "nature of sovereignty." If we are to understand how such naturalization has come about, however, it is not enough to apply ahistorical concepts; we must look back at the origins of these geopolitical discourses.

Geopolitics and "Natural Selection"

The "science" of geopolitics was rooted in the idea that national autonomy and control—that is, state sovereignty—were to be valued above all, and that to rely on the goodwill of others, or the "proper" functioning of international markets, was to court potential national disaster. Classical geopolitics saw the power, prosperity, and prospects of a state as fixed by geography and determined by inherent geographical features that could not be changed.[10] Territories could not be bought and sold; as parts of integral nation-states, they might be wrested or stolen in battle, but they were not for sale at any price.[11] Classical geopolitics was a product of its age, the Age of Imperialism and Social Darwinism, not the Age of Liberalism and Ecology. It is no coincidence that the best-known progenitors of geopolitics were citizens of those great powers—Britain, Germany, the United States—that sought to legitimate international expansion and control through naturalized ideological covers.

To give some of these scholars their due, not all regarded geography as fully binding on state autonomy and action. Halford Mackinder[12], a Briton, was initially less of a geopolitical determinist than Nicholas Spykman, an American.[13] World War II hardened both, inasmuch as Germany's efforts to expand appeared to vindicate Mackinder's theory about "Heartland" and "Rimland" powers (a dichotomy later picked up by Colin Gray as well).[14] And following World War II, a vulgar sort of geopolitical determinism came to dominate much realist theorizing as well as foreign policy analysis,[15] triggered to no small degree by George

Kennan's "Long Telegram."[16] Such determinism was routinely made part of every strategy and policy document to come out of the U.S. government. A typical assertion of this sort can be found in NSC-94, "The Position of the United States with Respect to the Philippines."

From the viewpoint of the USSR, the Philippine Islands could be the key to Soviet control of the Far East inasmuch as Soviet domination of these islands would, in all probability, be followed by the rapid disintegration of the entire structure of anti-Communist defenses in Southeast Asia and their offshore island chain, including Japan. Therefore, the situation in the Philippines cannot be viewed as a local problem, since Soviet domination over these islands would endanger the United States' military position in the Western Pacific and the Far East.[17]

More recently, Colin Gray, perhaps the last of the classical geopoliticians, argued that "because it is rooted in geopolitical soil, the character of a country's national security policy — as contrasted with the strategy and means of implementation — tends to show great continuity over time, although there can be an apparently cyclical pattern of change."[18] If Gray's claim is correct, the facts of national fate are written in Nature, and there can be no struggle against them. In his 1988 book, Gray simply takes boundaries — in this case, between the United States and the Soviet Union, which three years later would cease to exist — as given and as "natural" as the "geopolitical soil" in which they are drawn. (In his subsequent book, Gray has adopted a common practice of the 1990s, naturalizing culture as a product of geography, in order to warn that the United States, as a maritime power, must remain on guard against Russia, a "Heartland" power.[19])

The Age of Imperialism was also the Age of Social Darwinism, rooted in Charles Darwin's ideas about natural selection, but extended from individual organisms as members of species to states. As Agnew and Corbridge argue,

Naturalized geopolitics [from 1875 to 1945] had the following principal characteristics: a world divided into imperial and colonized peoples, states with "biological needs" for territory/resources and outlets for enterprise, a "closed" world in which one state's political-economic success was at another's expense . . . , and a world of fixed geographical attributes and environmental conditions that had predictable effects on a state's global status.[20]

According to German philosophers, states could be seen as "natural" organisms that passed through specific stages of life; hence, younger, more energetic states would succeed older, geriatric ones on the world stage. States must therefore continually seek advantage in order not to succumb prematurely to this cycle of Nature; as Simon Dalby puts it, "[S]tates were conceptualized in terms of organic entities with quasi-biological functioning. This was tied into Darwinian ideas of struggle producing progress. Thus, expansion was likened to growth and territorial expansion was ipso facto a good thing."[21] British and American geopoliticians had a somewhat different perspective, seeing progress tied to "mastery of the physical world" through science and technological innovation. But Nature was still heavily determining.

[B]efore the First World War, the current European geopolitical vision linked the success of European civilization to a combination of temperate climate and access to the sea. Temperate climate encouraged the inhabitants to struggle to overcome adversity without totally exhausting their energies, hence allowing progress and innovation to lead to social development. Access to the sea encouraged exploration, expansion and trade, and led ultimately to the conquest of the rest of the world.[22]

In this invocation of Nature (and nature), the geopoliticians transformed sovereignty from an attribute of a political community to a characteristic of a "natural" entity. Sovereignty was thereby naturalized.

Both perspectives—organic and innovative—also helped to legitimate imperial expansion, colonialism, and conquest as critical to the maintenance of national sovereignty. The "life cycle" argument demanded adequate access to the material resources and space necessary to maintain national vitality—hence, the German demand for colonies and, later, lebensraum. The "struggle to survive" required colonies as well as position to command the vital geographic features that would provide natural advantages to those who held them—hence, British garrisons from Gibraltar to Hong Kong, and the Panama Canal Zone and Pacific Islands under American suzerainty. Geopolitics could thus be cast as a "science," one well suited to the neomercantilism and gold standard of the late nineteenth and early twentieth centuries. One hundred years of industrialization of Europe provided the impetus to policies of development as well as territorial unification under the rubric of "nation." Each nation was the autochthonous offspring of the land where it lived—which created problems for those nations, such as the Germans, who were scattered throughout Central and Eastern Europe. In this way, national territory became not only sacred but "Natural." Only those within the natural borders of the nation-state could be mobilized to serve it and only those who were naturalized—that is, loyal to the nation-state—could be relied on. It is no accident that borders, so fluid during the age of sovereigns, became rigid, with passports required, during the age of sovereignty.

The ethnic cleansings and population transfers of the twentieth century illustrate this point. Woodrow Wilson's doctrine of national self-determination helped to further this process by extending the mythic principle of the organic nation to all those who could lay claim to such status. Where competing claims arose within specific territories, the strong tried to assimilate, eliminate, or expel the weaker.[23] We continue to observe this process in many places around the world, even as markets have rendered borders porous and control over them problematic.[24] Indeed, the "culture wars" that have spread to a number of industrialized countries over the past decade are as much about restoring the mental borders of the nation as they are about expelling those with alien ideas and identities.[25] Passports can be falsified; beliefs cannot.

The fundamental erosion of state sovereignty by markets is nowhere seen more clearly than in the realm of genetics, where the geopolitical discourse of states has been transformed into a "geopolitics of the body." The invocation of nature to demonstrate the superiority of human groups is not a new phenomenon, and can be traced at least as far back as the ancient Greeks. For the past century, Darwin's theories of evolution have been used to legitimate the genetic superiority of some races and nations over others. The latter form of naturalization was greatly delegitimated as a result of its application by various enthusiasts of eugenics, but it has reemerged in a somewhat different form over the past twenty years or so.[26] This new genre of naturalization—genetic determinism—has developed as both science and ideology, and its parallels to older geopolitical and organic theories of nation and nationalism are worth noting.

The contemporary scientific basis for genetic determinism is found in the various research efforts that seek to understand the genetic basis for various congenital diseases and inherited characteristics, culminating in the Human Genome Project.[27] The ideological manifestation, however, reflects an almost pure version of liberal methodological individualism in its framing: An individual's potential is almost wholly inherent in her or his genetic inheritance. Twin and sibling studies seem to suggest that society and environment are at best minor contributors to that potential, with the result that, in effect, one is already of the "elect" at birth (so one would do well to be careful in choosing one's parents).[28]

As is true with geography and the state, an individual's "natural" inheritance is critical to that person's development. But the ways in which this particular (and not terribly innovative) insight is being used politically are rather alarming.[29] In particular, genetic determinism is helping to reconstruct a vulgar form of Hobbesian genetic "war of all against all," in which the individual has no one to blame but herself for anything that might befall her in the marketplace of life. Inasmuch as the state has been banished from this realm (except as a declining source of research funds), there is no one to turn to for protection against predation by others with superior genetic endowments or with sufficient cash.[30] Another version of this ideology extrapolates natural inheritance back to race and ethnicity, arguing that society has no responsibility to redress historical inequities inasmuch as these are largely genetic in origin. Again, it is sink or swim in the genetic marketplace.

In this hyperliberal Nature, consequently, a new form of sovereignty accrues to the individual. Here, control is exercised by those with good genes — which are scarce — or who have the wealth to acquire them. Because wealth in the market is power, it is also the key to preventing oneself from being contaminated by "bad" genes carried by the poor, the ill, the defective, or the alien. Such quality, of course, carries on into one's offspring. As with classical geopolitics, the naturalized discourse of genetics follows the dominant ideology of the day and, in some of its more extreme expressions, involves an almost complete transfer of the "natural rights" associated with sovereignty from the state to the individual. The result is that sovereignty devolves to the individual through genetics; as an attribute of the state, it is disintegrating in both material and ideational terms.

Sovereignty, Property, Interdependence

What, then, can we mean by *sovereignty*? Although the term continues to be the focus of vociferous controversy, especially as it appears to many to be "eroding," I prefer to follow Nicholas Onuf's lead here and conceptualize it as a property of liberalism.[31] He cites C. B. Macpherson's description: "The individual is free inasmuch as he is the proprietor of his person and capacities. The human essence is freedom from dependence on the will of others, and freedom is a function of possession. Society has a lot of free individuals related to each other as proprietors of their own capacities and of what they have acquired by their exercise."[32] Onuf then points out that "[s]tates are granted just those properties that liberalism grants to individuals,"[33] among which are real estate, or property. (This is easier to understand if we recall that, for the original sovereigns of the seventeenth century, states were property).[34] In a liberal system, individuals holding property are entitled to use it in any fashion except that which is deemed harmful to the interests or welfare of the community,[35] and, indeed, this is precisely the wording of Principle 21 of the Stockholm Declaration: States have the right to exploit their own resources so long as this does not have an impact on the sovereignty of other states by constituting an illegal intrusion into the space of jurisdiction of other states.[36] What this implies, therefore, is that boundaries are important. Inside the boundaries of property the state, like the individual landowner, is free from "dependence on the will of others"; outside the boundaries, it is not — at least, in theory.

Practice is quite different. The individual owner finds her sovereignty over real property not only hedged about with restrictions but also subject to frequent intrusion due to others' wills. Indeed, the state has the prerogative of violating the sovereignty of private property in any number of settings, ranging from suspicions about the commission of crimes on property to the taking of property in the greater social "interest" — subject to just compensation (markets are involved only so far as setting the "value" of the property is concerned). The property owner has little recourse in these situations, except to the courts. Such is the power of law. The state, by contrast, has freed itself from such legal niceties through resort to the fiction of international "anarchy" and "self-help," the doctrine we know as *realism*. That doctrine allows the state to physically resist violations of its property, on the one hand, while declaring a national "interest" in violating the property of other states. This legitimates its right to transgress boundaries, notwithstanding the Stockholm Declaration and other international laws of a similar bent.

For reasons that are beyond the scope of this chapter to investigate, such egregious violations of territorial property and sovereignty are increasingly frowned on. This has not, however, led to a diminution in violations of sovereignty. Recall, for example, that the distribution of resources among states is uneven, a condition often blamed on nature and geography, with the result that one state finds itself needing to interact or engage in exchange with another. This state of affairs is generally characterized as *ecological interdependence*,[37] a situation whereby state borders, characterized as "natural" under sovereignty and anarchy, fail to correspond to those of physical and biological nature. It is this tension between territorial sovereignty and the sovereignty of nature that sets up the basis for problems such as "water wars" in the first place. Below, I will examine the concept of ecological interdependence more closely; here, I only point out that, while this notion is often taken to describe a physical phenomenon—the existence of ecological phenomena crossing, or ecosystems extending across, national borders—it actually serves to obscure relations of domination between the states in question.[38] Inasmuch as rights of property—state sovereignty—reify the control of a resource, neighboring states that may lack access to or control of the same resource find themselves in a condition of relative powerlessness with respect to it. Their only recourse in such a situation is to physically capture the resource or to purchase property rights to it and thereby "legally" control it.

We now begin to see how and why water wars and water markets can be so easily juxtaposed in the language and reasoning of liberalism and neoclassical economics. A water war is simply the international equivalent of an unjust "taking" without the constitutional gobbledygook. Water must belong to someone, presumably to the country within which it is physically located. In the absence of a legal means of transferring property rights, what is a thirsty state to do except, presumably, grab the stuff? Water markets thus become the equivalent of a taking with just compensation (although the definition of *just* in this context is a function of markets rather than meanings). Sovereignty, like property, can be bought and sold. Whether water markets will eliminate putative water wars is an entirely separate matter. Creating open, transborder markets in water will not necessarily lead to "water peace." It could mean, instead, that the highest bidder wins the water and the losers get angry.[39] Indeed, it will probably be more "efficient" and make more economic sense for Palestine to sell its water to Tel Aviv rather than use it for West Bank agriculture. So the water will flow across borders and, once again, become scarce on one side of the border and plentiful on the other, thanks to the control made possible through markets rather than military occupation. It remains to be seen how this will sit in Hebron and Nablus.

Scarcity and the "Limits to Growth"

There is no need here to recount in detail the geneaology of "scarcity" as a concept;[40] suffice it to say that it is central to the theory and practice of neoclassical economics. Instead, let us consider the relationship between scarcity and boundaries or, rather, between the conditions that differentiate between absolute and relative scarcity, and the politics associated with both. Continuing the argument made above, scarcity is clearly a product of human institutions: of control, of ownership, of property, of sovereignty, of markets. Economists often tell us that absolute scarcity does not—indeed, cannot— occur if and when markets are operating properly: All scarcity is relative. In an "efficient" market, free of political intervention, should the supply of some good run low, its price will rise and people will seek less expensive substitutes.[41] Doomsayers such as Malthus and the Ehrlichs have thus been criticized for ignoring the rules of supply and demand and, up to a point, the critics are correct. When we insert boundaries into the equation, however, it turns out that the doomsayers do have something to say.

Malthus was a prophet of absolute scarcity.[42] He argued that geometrical population growth would eventually outstrip the arithmetical growth of agricultural production. This would result in circumstances in which food would run short in absolute terms, leading to widespread starvation and death. His analysis has been—and continues to be— criticized for not taking into account either basic economics or technological innovation, but these criticisms are not very fair. As a cleric, Malthus was undoubtedly more interested in distribution than in markets or capital and, from a strictly ecosystemic perspective, he was right:

When food runs short, populations crash. And most populations have recourse neither to markets nor the means of moving food from one place to another. They can move, but if all neighboring niches are occupied, the game is up.[43]

A similar notion of absolute scarcity was promulgated several centuries later by Dennis and Donella Meadows and their colleagues at MIT.[44] They concluded that, given then-current trends in nonrenewable resource production, reserves, and consumption, and barring unforeseen circumstances or discoveries, the world would run short of various critical materials sometime during the twenty-first century. The Meadowses and their colleagues were harshly taken to task for ignoring the same factors as Malthus. To the satisfaction of many, they were soon "proved wrong" by events. Even today, economists still find pleasure in pointing this out.

What crude Malthusianism and Meadowsism both disregarded was the matter of distribution of resources—that is, for whom would food and minerals be scarce? And why would it matter? Certainly, it is by no means clear that the depletion of global chromium supplies would matter as much to Chinese peasants as to Cambridge academics. For this error, in any event, Meadows and his colleagues should be forgiven; economists tend to dismiss the same point, too, regarding distribution as an "exogenous" problem and one that, in any event, can be addressed by economic growth. Their supply and demand curves do no more than illustrate the premise that, if scarcity drives prices to rise too high relative to demand, markets will be out of equilibrium and no one will buy. Eventually, sellers will have to lower their prices, and buyers will be able to eat again. The reality is slightly more complicated, inasmuch as even properly functioning markets can foster maldistribution and relative scarcity; as Amartya Sen has pointed out, not everyone starves during a famine—indeed, food is often quite plentiful.[45] What crude market analyses don't take into account is that, even at market equilibrium, there may be those for whom prices are still too high. Those who have money can afford to buy; those that do not, starve. Scarcity is only relative in this instance, but some people (and countries) do go hungry.

In other words, relative scarcity is also a condition of boundaries, in this instance political, cultural, or class ones. In some instances, these lines are found between the physical personas of individuals: I am of one caste (class, ethnic group, religion); you are of another. My money and food are mine (or my group's), not yours (or your group's). In other cases, the lines are drawn between countries: This land (and water) is ours, not yours. At both extremes, the money, food, land, water, and other resources must be kept inside that boundary in order to maintain individual and collective integrity, identity, and sovereignty. The resources must remain sovereign property. And this goes beyond simple material possession, too.

If I give you my money, so that you can buy food, I will have less and will not be able to live the way I am accustomed. This will lessen me. In doing this, I will also acknowledge a relationship with you that infringes on me and even acknowledges my obligations to you. If I do this, then I will not be who I have been because I will have yielded some of my autonomy to you. Moreover, because you have no money, you cannot buy from me; and because I have as much as I need, I don't have to buy from you. Hence, I can remain sovereign and strong.[46]

The transfer of anything across a boundary—whether physical, political, social, cultural, or economic—serves to acknowledge the existence of an Other and, in establishing a relationship, thereby lessens sovereignty—that is, control and autonomy, whether individual or collective—in an absolute sense. It also creates social relations between and among actors that have little or nothing to do with markets or anarchy.[47] Unless one asserts total control over that something—which implies assimilating or eliminating the Other—sovereignty can never be restored to its idealized form as hypothesized in the State of Nature.

In other words, sovereignty, whether individual or national, is about exclusion and autonomy, both physically and cognitively. This helps explain why uneven distribution is so important in international politics: It helps to perpetuate the hierarchy of power that, notwithstanding the fiction of functionally equivalent units, is central to international politics and markets. That was the purpose of the princes' agreement at Westphalia; that is the point of the reification of methodological individualism today. Inside my boundary, I/we can act as we wish; outside of it, I/we cannot. By redrawing or even, in some circumstances, abolishing lines, we could change this premise, but that would mean sharing what we have with others and having less for ourselves. It would change us. And this is why autarchy—the abolition of relative scarcity through a redrawing of lines of control—was long the dream of realists and policymakers. But autarchy is economically costly. Interdependence, whether economic, ecological, or military, is one way of finessing this problem without redrawing boundaries, yielding sovereignty, or shifting identity.

Embedded Liberalism and Interdependence

Interdependence, as the term is commonly used in international politics, is a consequence of U.S. geopolitical strategy during the Cold War, and not Nature per se. Following World War II, the geopolitical discourses of Mackinder, Spykman, and others were transmuted into the containment policy attributed to George Kennan and formalized in NSC-68 and other realist texts. There was, however, a problem. The neomercantilistic geopolitics of the late nineteenth and early twentieth century states were unsuited for the postwar world.[48] This was because neomercantilism regarded the nation-state—or, perhaps, empire—as the natural unit of economic-military action. But "national capitalism," as Fred Block called it[49] and which was practiced during the interwar period, would result in restricted and too-small markets. This would depress economic growth, risk a return to Depression conditions, and, perhaps, trigger World War III. Inasmuch as the liberalization project of American postwar planners posited an economic space larger than the national territorial space—in order to make sure that goods crossed borders so that soldiers would not—a new unit of economic activity had to be developed: the "Free World."

Originally, of course, this space was not called the Free World, inasmuch as the Soviets had not yet been definitively tagged as the new enemy.[50] Harry Truman's felicitous allusion to "free peoples everywhere" provided the label; the imperialism of the dollar and the fear of Reds did the rest. Within the borders of the Free World, all states were united in pursuit of common goals based on the human propensity to "truck and barter"; outside of those borders were those whose behavior was "unnatural." As films such as the 1956 version of *Invasion of the Body Snatchers* suggested, communism was a pathology of nature, not an ideology of men; it took you over, you did not take it on.[51] Keeping the enemy out and contained meant, therefore, not only imposing secure boundaries around the world but also imposing limits on one's own self. The domino theory was not only about the fall of states; the rupture of containment would breach the self as well. The success of the Free World thus depended on extending boundaries around a natural community that had not, heretofore, existed.[52] But in order to maintain its sovereignty and autonomy, this natural community had to be juxtaposed against another. Thus, on one side of the boundary of containment was to be found a unit—the Free World—whose sovereignty depended upon keeping out the influences of a unit on the other—the bloc; in other words, the Free World could not exist without the corresponding "Unfree World."

Within the borders of the Free World, however, there remained a problem: The maintenance of state sovereignty and autonomy—previously regarded the natural order of things—threatened to undermine the integrity of the whole. This was especially difficult from the American point of view, as illustrated in the famous confrontation between so-called isolationists and internationalists.[53] The solution to this dilemma was what John Ruggie calls "embedded liberalism," a form of multilateral economic nationalism.[54] Inside the boundaries of the Free World, states were granted the right to manage their national economies, but only so long as they agreed to move toward and, eventually, adopt the tenets of an internationalized liberalism. With respect to the area outside the boundaries, however, the Free World would, to the extent possible, be neomercantilistic.

In the long run, this proved to be a difficult balancing act. To institutionalize liberalism throughout the Free World required that states yield up sovereignty: in the name of defense and free markets, upward to the system; in the name of democracy and prosperity, downward to the individual, the "natural" unit of interaction in the market and the final holder of human rights. In particular, to fully carry through this shift meant that the state would have to yield up its sovereign prerogatives and insularity to the market, which would breach its borders with flows of raw materials, manufactures, technology, ideas, culture, capital, and even labor. The United States, as the center of this global market system, would remain technologically dominant and thereby retain its competitive edge and autonomy to manage the Free World (although this is not what has actually come to pass). In theory, all barriers would have to fall to fully realize the potential of liberalism; in practice, there was strong resistance, first in Europe (as evidenced, for example, in the establishment of the European Common Market) and later elsewhere.

Conditions within the Free World economic area did not constitute "interdependence" as that concept came to be defined during the 1970s; rather, they were a manifestation of American material domination and Gramscian hegemony.[56] The geopolitical discourse later developed to account for the halcyon conditions of the 1950s and 1960s invoked "hegemonic stability theory."[57] It is helpful to jump ahead of our story for a moment to consider the "double hermeneutic" of this theory. Originally formulated as a term of socialist political opprobrium, between about 1975 and 1985, roughly corresponding to the period of generalized worries about American decline, *hegemony* was naturalized and given positive attributes. A *hegemon* was a state destined for power, through the cycles of history and global political economy, which took on the burdens of global economic and political management under anarchy,[58] through the establishment of international regimes in which it remained dominant. In so doing, not only were its own interests served, first and foremost, but so were those of other countries, free-riding on the hegemon. Under the skillful hands of non-Marxist students of international political economy, domination was thereby transmuted into a type of stewardship—with scholars such as Robert O. Keohane subsequently worrying about what would happen "after hegemony." Free riders failed to appreciate these benefits,[59] and their reluctance to share the burden played a major role in the political disorder of the 1970s and the renewed Soviet threat during the 1980s. Ungrateful wretches! The United States—the defender of naturally free men [*sic*] and markets—had only the best interests of its allies in mind when it manipulated or coerced them.

As with hegemonic stability theory, interdependence theory was also an academic product and geopolitical discourse of the times, rather than an objective description of reality. *Interdependence theory* tried, on the one hand, to account for what appeared to be the end of American dominance of the Free World after the dollar devaluations and end of convertibility to gold and the oil embargo of the early 1970s. On the other, it was also used to justify certain policies and actions that might otherwise be politically unpopular at home and abroad. But, whereas hegemonic theory derived in large part from realism, interdependence theory was liberal in origin (although, as Robert Keohane demonstrated, the two were perfectly compatible). Indeed, in 1977, Keohane and his collaborator Joseph Nye proposed that "interdependence in world politics refers to situations characterized by reciprocal effects among countries or actors in different countries."[60] What did they mean by "reciprocal effects"? Writing in the late 1970s, in aftermath of the first runup in oil prices, they clearly had the distribution problem in mind. The United States no longer owned enough oil under its private sovereignty, within its national boundaries, at an acceptable price; others owned too much. (No one paid any attention to the reciprocal effects on others of too much oil at too low a price, a condition in the early 1960s that led to the establishment of OPEC).[61] For the United States, the "reciprocal effect" was, therefore, primarily a problem of domestic politics, a disgruntled electorate forced to queue and shell out twice or thrice what they had paid previously for a gallon of gas. A number of analysts and policymakers — including Henry Kissinger — proposed taking the oil back, inasmuch as it "belonged" to the United States through American oil companies. In time, cooler heads prevailed, but we were no longer the same nation we had been prior to 1973.

Keohane and Nye further distinguished between "sensitivity" and "vulnerability." If you are sensitive, they said, you feel the impacts of an action by another but you can recover — that is, you can eliminate the temporary infringement on your autonomy, sovereignty, and identity imposed by others. Then it is back to business as usual. If you are vulnerable, however, you feel the impacts and cannot recover — your autonomy and sovereignty have been breached for good and you have been changed. You are now constrained by a new relationship with an Other that, as much as you might dislike it, has served to establish a new element of your identity in terms of that Other.[62] To be sure, such a merging of identities can exacerbate the sense of those differences that remain, a phenomenon most evident today in the clash between religious fundamentalisms and secularisms. But the two identities are mutually constitutive, not oppositional; each cannot exist as an identity without being bound to the other. By contrast, interdependence connotes transactions or exchange across boundaries and, consequently, some degree of separateness and mutual alienation, as in a market setting. To speak, then, of interdependence is to make an effort to restore the breached boundaries between oneself and the Other, rather than to adapt to this new condition of mutual constitutiveness.

Whatever the merits of the distinction made by Keohane and Nye, both sensitivity and vulnerability posit impacts across borders, infringements on autonomy, and reductions in sovereignty. To eliminate either type of intrusion, adjustments must occur inside the boundaries—in the realm of domestic politics—so that the boundaries can be restored. Hence the contradiction: Interdependence is acceptable if it allows us to maintain the boundaries and our national sovereignty, unacceptable if it does not. At best, this is a word game; at worst, a form of false consciousness that serves to perpetuate domination. Or, as Edward Said has written, "This universal practice of designating in one's mind a familiar space which is 'ours' and an unfamiliar space beyond 'ours' which is theirs is a way of making geographical distinctions that can be entirely arbitrary."[63] The lines are thus drawn around "natural" communities—that is, ones united by characteristics or culture whose origins are perhaps lost in history, mutually alienated from one another by virtues of these differences.

Keohane and Nye acknowledged (but deplored) the ideological content of the concept when they wrote,

Political leaders often use interdependence rhetoric to portray interdependence as a natural necessity, as a fact to which policy (and domestic interest groups) must adjust, rather than as a situation partially created by policy itself.... For those who wish the United States to retain world leadership, interdependence has become part of the new rhetoric, to be used against both economic nationalism at home and assertive challenges abroad.[64]

Paradoxically, perhaps, the rhetoric and theory of interdependence was also intended to reinforce boundaries without sealing them off as those at home and abroad might wish. The unit of analysis remained the sovereign state; the impacts posited by Keohane and Nye impinged on states rather than individuals, classes, or other groupings; the responses would be taken by policymakers in the "national interest."

Consequently, political policies took the form of state-led actions. The oil embargo of 1973 and subsequent price hikes were presented as being directed against countries, with their citizens represented as homogenous entities, even though there were differential effects within and across the target states and on people. U.S. politicians fulminated about OPEC taking "our oil" and infringing on "our sovereignty," even though the oil was "owned" by multinational corporations and stockholders, based in the industrialized states, who profited handsomely as prices rose. And state-led policies to redress these conditions, such as President Nixon's ill-fated Project Independence and President Carter's Synfuels Corporation—both subjects of fierce domestic attack for their intervention into markets—were presented as schemes to reduce all Americans' reliance not on oil, per se, but on that oil over which we could not exercise control.

For other countries, the rhetoric of interdependence did not signal the equalization of power relations among allies so much as a U.S. effort to rationalize anew its "natural" leadership of the Free World. This coincided—not accidentally—with the first stirrings of the renewed Cold War, marked by the rise of the second Committee on the Present Danger, the collapse of détente, and the political and academic reification of hegemonic stability theory. The erosion of boundaries around the world led to the effort to reinforce them in Central Europe and, eventually, elsewhere. By the end of the decade, the discourse of economic interdependence had dissolved, to be replaced by reassertions of sovereignty and autonomy under Jimmy Carter and Ronald Reagan, and legitimation of their policies through the newly created discourse of benevolent but vigilant hegemony. What the U.S. military buildup did not do, however, was to get rid of those borders; instead, it reinforced them.

Limits to Sustainability

The Stockholm Declaration was quite explicit about the environmental rights and responsibilities of states. States were sovereign entities and could not engage in activities that negatively affected the sovereignty—the property rights—of other states where resources and environment were concerned. What, in practice, did this mean? First, states possessed the absolute right to do whatever they wished with the natural resources located within their boundaries. Second, states were absolutely enjoined from doing anything with their resources that would somehow affect the sovereignty of any other state. Third, because the environment was not subject to these imagined boundaries, states were admonished to protect it within the conditions implied by the first two principles. In other words, the Stockholm Declaration reiterated the absolute impermeability of the boundaries of states as a condition of protecting the environment.

While there were (and are) any number of contradictions embedded in these principles, three stand out. First, in spite of long-standing evidence that nature "respects no borders," the state was once again reified as the sole appropriate agent of control, management, and development where environment was concerned. Second, the Stockholm Declaration granted to the state and its agents the absolute right to discipline nonstate actors when degradation of environment and natural resources was involved, even to the point of appropriation through coercion.[65] (That this might lead not only to further degradation, but also to ever-greater concentrations of power in the hands of those who encouraged degradation, was not immediately obvious and certainly not an argument that diplomats and policymakers wanted to hear.) And, third, it had the effect of reinforcing the natural separation of political units rather than fostering mutual and respectful relationships between them.

The reification of borders and sovereignty as Natural has had two difficult-to-reconcile consequences where environmental protection is concerned. On the one hand, as is often recognized, it fragments jurisdictions that might better be treated as single units. Thus, for example, sulfur dioxide emissions from power plants in the midwestern United States are a domestic regulatory problem when they rain out in the Northeast, but a "transborder" problem when they rain out a few miles further on, in Canada. On the other hand, such reification mandates "cooperation" among states, which seek to protect their sovereign economic prerogatives even as they attempt to address such transboundary issues. Such cooperation appears "difficult" and "unnatural" because it goes against the grain of anarchy and requires that states yield a measure of their domestic sovereignty in order to address the consequences of infringements on sovereignty enjoined by the Stockholm Declaration and periodically reiterated since then.

Both the rhetorics of ecological interdependence and "sustainable development" can, consequently, be interpreted as responses to the ontological difficulties posed by Stockholm; more to the point, they can both be understood as descendants of earlier geopolitical discourses. In 1987, the World Commission on Environment and Development, also called the Brundtland Commission, proposed that "the Earth is one but the world is not," and argued that the "world in which human activities and their effects were neatly compartmentalized within nations" had begun to disappear.[66] The commission further suggested "that the distribution of power and influence within society [*sic*; which society is not made clear] lies at the heart of most environmental and development challenges."[67] In other words, according to the commission, even though the sources of the problems of environment and development were to be found at both supra- and subnational levels, sustainability would nonetheless depend on states acting within and across boundaries, even when it was not evidently in their interest, ability, or willingness to do so.[68]

A more fundamental difficulty, however, has to do with who decides what is sustainable. From a crude ecological perspective, sustainability is frequently defined as a rough equivalence between inputs and outputs. This, it is widely assumed, is equivalent to conditions in nature that are naturally "balanced," even as ecologists warn us that ecosystemic balance and stability do not, for the most part, exist. To exceed this balanced condition for too long, or by too much, will run up against natural limits. From an economic (and nineteenth-century organic) perspective, however, growth is "natural." Consequently, sustainability rests on the accumulation of capital—and technology—that can be used to substitute for depleted resources or to rise above such limits. In its famous definition of the "solution," the Brundtland Commission finessed this problem by including both conceptions and ignoring the contradictions between them.

Humanity has the ability to make development sustainable—to ensure that it meets the needs of the present without compromising the ability of future generations to meet their own needs. The concept of sustainable development does imply limits—not absolute limits but limitations imposed by the present state of technology and social organization on environmental resources and by the ability of the biosphere to absorb the effects of human activities.[69]

The naturalization of biospheric limits, whatever they might be, are balanced in this definition by the naturalization of technology and social organization that know no limits. These, in turn, are to be discovered by recourse to markets that will, somehow, find a balance between needs and growth.[70]

The second response to the ontological problem of boundaries and sovereignty — the conflation of statist realism and methodological individualism discussed earlier in this chapter — is no more helpful. States fight over natural resources such as water because, as Peter Gleick has put it, "of its scarcity, the extent to which the supply is shared by more than one region or state, the relative power of the basin states, and the ease of access to alternative freshwater sources."[71] In other words, the "cause" of water wars is distribution, not supply per se. Prevention of conflicts is to be found in "formal political agreements" that will address the distribution problem, through the allocation of property rights to water and the appropriate pricing of water in markets among "rational" users.[72] Voilà! The naturalized factors that lead to interstate competition and war in the first place are subsumed by processes of economic competition and exchange in naturalized markets. These, it is presumed, will produce more efficient and, therefore, more peaceful outcomes. As I suggested earlier, however, the problem of unequal distribution will not go away; it will simply be shifted onto those who lack the power to make trouble. Sustainability will thus come to be defined not by the justice of distribution but by the judgment of markets; ecological interdependence will fall before wealth rather than force of arms, as the rich disempower the poor. Boundaries will be naturalized once again and sovereignty restored to its rightful place in the hierarchy of Nature.

What Does It Mean to Be "Natural" Where States Are Concerned?

It has been said that, for much of human history, nature was sovereign. Nature "made" the rules and humans were obliged to observe them or die. But human beings escaped from that (first) nature long ago, replacing it with one that was, for the most part, socially constructed and regulated (second Nature). To be sure, droughts, floods, pestilence, locusts, earthquakes, volcanoes, blizzards, and asteroids remind us that we do not control all biogeophysical processes. Still, that we are almost six billion strong clearly suggests that nature's sovereignty is, for better or worse, quite limited in some respects. I do not mean to suggest that we are unconstrained by "limits to growth," whatever those might turn out to be. I do mean to suggest that nature's "sovereignty" is not as limiting as many make it out to be, especially as framed through contemporary geopolitical discourses of rich and poor. I also mean to suggest that the "seamless web of global ecological interdependence" is, to no one's great surprise, more of a social construct than a biological or geophysical "fact."

At some level, we do all share the same atmosphere, climate system, and hydrological cycle; at the same time, we remain separated by all kinds of boundaries, not the least of which is that demarcating power from weakness. Indeed, it is a sign of power to be able to draw such boundaries and make them stick. And if there is anything the powerful do not want, it is to be burdened with the demands of the poor and weak (as is only too evident today within the United States). Hence, while the invocation of "interdependence" is a commonplace, it is virtually always viewed as a cost to us, our sovereignty, and our autonomy. In other words, not only is the tradeoff between "state sovereignty" and "ecological interdependence" a false one, as Karen Litfin has suggested;[73] both concepts are also discursive power plays that fall back on naturalized concepts in order to protect privilege. Those who do not want to alter their ways of doing things can call on *national* sovereignty for protection; those who want others to change their ways of doing things can call on *natural* sovereignty. The central issue here appears to be about control over outcomes; the central controversy is, control by whom? Either way, someone seeks to control someone else and naturalize the process as a means of justifying it.

We might do better to denaturalize the ways in which we think about the world and relate to each other, state to state, person to person, people to nature. Nature has been around for billions of years, and the state has been with us for centuries. Still, the state is not natural and the Hobbesian/Lockean States of Nature are political fiction. We need to acknowledge the social character of power, control, domination, competition, and, indeed, war, rather than treating them as immutable and inevitable. We need to recognize that, while we are separate beings we are, at the same time, mutually constituted by each Other. If we don't come to grips with this recognition, we will continue to draw new lines. Then, the mantra of the next century will probably be a very hostile "Get out of my space, man!"

Notes

1. World Bank, "Flowing Uphill," *Economist* August 12, 1995): 36.

2. To name just a few: Natasha Beschorner, *Water and Instability in the Middle East* (London: Brassey's for the International Institute for Strategic Studies, 1992); Daniel Hillel, *Rivers of Eden: The Struggle for Water and the Quest for Peace in the Middle East* (New York : Oxford University Press, 1994); J. Isaac and H. Shuval, eds., *Water and Peace in the Middle East: Proceedings of the First Israeli-Palestinian International Academic Conference on Water, Zurich, Switzerland, 10–13 December 1992* (Amsterdam: Elsevier, 1994); Elisha Kally with Gideon Fishelson, *Water and Peace: Water resources and the Arab-Israeli Peace Process* (Westport, Conn.: Praeger, 1993); Masahiro Murakami, *Managing Water for Peace in the Middle East: Alternative Strategies* (Tokyo: United Nations University Press, 1995); Joyce Starr and Daniel C. Stoll, eds., *The Politics of Scarcity: Water in the Middle East* (Boulder, Colo.: Westview Press, 1988); Joyce Starr, *Covenant over Middle Eastern Waters: Key to World Survival* (New York: H. Holt, 1995); Aaron T. Wolf, *Hydropolitics along the Jordan River: Scarce Water and Its Impact on the Arab-Israeli Conflict* (Tokyo: United Nations University Press, 1995); Peter Gleick, "Water, War, and Peace in the Middle East," *Environment* 36, no. 3 (April 1994): 6–15, 35–42; Miriam Lowi, "West Bank Water Resources and the Resolution of Conflict in the Middle East," Occasional Paper Series of the Project on Environmental Change and Acute Conflict no. 1 (September 1992): 29–60; Joyce Starr, "Water Wars," *Foreign Policy* 82 (Spring 1991): 17–30; John Bulloch and Adel Dawish, *Water Wars: Coming Conflicts in the Middle East* (New York: Victor Gollancz, 1993).

3. "Ships are but boards, sailors but men: there be land-rats and water-rats, land-thieves and water thieves." William Shakespeare, *The Merchant of Venice,* act 1, sc. 3.

4. World Bank, "Flowing Uphill."

5. "Global deficiencies and degradation of natural resources, both renewable and nonrenewable, coupled with the uneven distribution of these raw materials, can lead to unlikely — and thus unstable — alliances, to national rivalries, and, of course, to war." Arthur H. Westing, ed., "Introduction," *Global Resources and International Conflict* (Oxford: Oxford University Press, 1986).

6. I use the term *State of Nature* to denote that semimythical, antedeluvian human condition written about by Hobbes, Locke and others, and attributed by international relations scholars to states operating under conditions of anarchy. I use the term *state of nature* to denote that contemporary condition of being constrained by the biogeophysical characteristics of the material world.

7. Ronnie D. Lipschutz, "Strategic Insecurity: Putting the Pieces Back Together in the Middle East," in: Harry Kreisler, ed., *Confrontation in the Gulf*

(Berkeley: Institute of International Studies, University of California Press, 1992): 113–126; Ronnie D. Lipschutz, "Raw Materials, Finished Ideals: Strategic Raw Materials and the Geopolitical Economy of U.S. Foreign Policy," in Martha L. Cottam and Chih-yu Shih, eds., *Contending Dramas — A Cognitive Approach to International Organizations* (New York: Praeger): 101–26.

8. Ronnie D. Lipschutz, *When Nations Clash — Raw Materials, Ideology and Foreign Policy* (New York: Ballinger/Harper and Row, 1989); Lipschutz, "Strategic Insecurity"; Lipschutz, "Raw Materials, Finished Ideals."

9. John Agnew and Stuart Corbridge, *Mastering Space: Hegemony, Territory and International Political Economy* (New York: Routledge, 1995).

10. As opposed to political geography, which studies the "relationship between geographical factors and political entities" (Hans W. Weigert, et al., *Principles of Political Geography* [New York: Appleton-Century-Crofts, 1957]). Geography can be changed, of course, as evidenced, for example, by the case of the Panama Canal. Oddly, perhaps, the canal served to enhance American power — it was now possible for the Navy to move from one ocean to the other more quickly — while also exacerbating vulnerability — any other power gaining access to the canal could now threaten the opposite U.S. coast more quickly.

11. The best example of this was the struggle over Alsace-Lorraine between France and Germany. Only the shedding of the nation's blood could redeem the lost pieces of the organic nation-state. For an historical perspective, see Norbert Elias, trans. Edmund Jephcott, *The Civilizing Process — State Formation and Civilization* (Oxford: Basil Blackwell, 1994).

12. Halford J. Mackinder, *Democratic Ideals and Reality* (New York: Norton, 1919/1962).

13. Nicholas J. Spykman, H. R. Nicholl, ed., *The Geography of the Peace* (New York: Harcourt Brace, 1944).

14. The theory was: "Who rules East Europe commands the Heartland; Who rules the Heartland commands the World-Island; Who rules the World-Island commands the World," (Mackinder, *Democratic Ideals*, 150). Halford J. Mackinder, "The Round World and the Winning of the Peace," *Foreign Affairs* (July 1943).

15. Lipschutz, *When Nations Clash*.

16. John Lewis Gaddis, *Strategies of Containment* (Oxford: Oxford University Press, 1982).

17. Cited in Lipschutz, *When Nations Clash*, 103.

18. Colin S. Gray, *The Geopolitics of Super Power* (Lexington: University Press of Kentucky, 1988: 15.

19. Colin S. Gray, *War, Peace, and Victory — Strategy and Statecraft for the Next Century* (New York: Simon & Schuster, 1990). Interestingly, culture has become the most recent refuge for many of those international relations

scholars who are unable to account for the vagaries of world politics; see, e.g., Samuel P. Huntington, *The Clash of Civilizations and the Remaking of World Order* (New York: Simon and Schuster, 1996); Francis Fukuyama, *Trust — The Social Virtues and the Creation of Prosperity* (New York: Free Press, 1995).

20. Agnew and Corbridge, *Mastering Space*, 57.

21. Simon Dalby, *Creating the Second Cold War — The Discourses of Politics* (London/New York: Pinter Guilford, 1990): 35. As noted below, this discourse has now been extended to the sovereign individual.

22. Dalby, *Creating the Second Cold War*, 35.

23. Ronnie D. Lipschutz, "(B)orders and (Dis)Orders: Sources and Sinks of Moral Authority in International Relations and Global Politics." Paper prepared for the International Studies Association Annual Meeting, March 18–22, 1997, Toronto, Canada.

24. See, for example, Susan Strange, *The Retreat of the State* (Cambridge: Cambridge University Press, 1996).

25. Ronnie D. Lipschutz, "From Culture Wars to Shooting Wars: Globalization and Cultural Conflict in the United States," in Beverly Crawford and Ronnie D. Lipschutz, eds., *The Political Economy of Cultural Conflict* (Berkeley: Institute of International and Area Studies, University of California Press, 1998).

26. For example, in Richard Herrenstein and Charles Murray, *The Bell Curve* (New York: Basic Books, 1994).

27. See, e.g, Lois Wingerson, *Mapping Our Genes — The Genome Project and the Future of Medicine* (New York: Plume, 1991).

28. Dahlem Workshop, *What Are the Mechanisms Mediating the Genetic and Environmental Determinants of Behavior? Twins as a Tool of Behavioral Genetics* (Chichester/New York: Wiley, 1993).

29. Some writers, such as Richard Dawkins, *The Selfish Gene* (New York: Oxford University Press, 1989) have gone so far as to argue that the appropriate unit of competition and survival is the individual "gene," and that humans (and, presumably, other species) are only containers for them. Of course, by that argument, bacteria and viruses are probably "bound to win."

30. Charles J. Hanley, "Blood Money," *San Francisco Examiner*, April 21, 1996.

31. Nicholas Onuf, *World of Our Making — Rules and Rule in Social Theory and International Relations* (Columbia: University of South Carolina Press, 1989).

32. Onuf, *World of Our Making*, 165, quoting C. B. Mcpherson, *The Political Theory of Possessive Individualism* (Oxford: Oxford University Press, 1962); 3.

33. Onuf, *World of Our Making*, 166.

34. Elias, *The Civilizing Process.*

35. See, e.g., Gary Libecap, *Contracting for Property Rights* (Cambridge: Cambridge University Press, 1989); John G. Ruggie, "Territoriality and Beyond: Problematizing Modernity in International Relations," *International Organization* 47, no. 1 (Winter 1993): 139–74.

36. Principle 21 abjures states to recognize the "responsibility to ensure that activities within their jurisdiction or control do not cause damage to the environment of other States or of areas beyond their national jurisdiction." Cited in Ved. P. Nanda, *International Environmental Law and Policy* (Irvington-on-Hudson, N.Y.: Transnational Publishers, 1995): 86.

37. Ronnie D. Lipschutz and Ken Conca, "The Implications of Global Ecological Interdependence," in Ronnie D. Lipschutz and Ken Conca eds., *The State and Social Power in Global Environmental Politics* (New York: Columbia University Press, 1993): 327–43.

38. Do we ever speak of "ecological interdependence" with, say, our children, spouses, significant others, or between California and Nevada?

39. How else to account for the difficulties in establishing water markets in California? Farmers seem resistant to any taking at any price, just or not, and with good reason. Selling water might make you richer, but a farmer will no longer be a farmer.

40. Even Thucydides mentions it; see *The Peloponnesian War* (Harmondsworth, U.K.: Penguin, 1954): 109. See also Simon Dalby, "Neo-Malthusianism in Contemporary Geopolitical Discourse: Kaplan, Kennedy and New Global Threats." Paper prepared for presentation to a panel on Discourse, Geography, and Interpretation, at annual meeting of the International Studies Association, Chicago, February 1995.

41. Of course, renewable resources are not subject to this particular economic logic, inasmuch as their flows are large and their stocks small or nonexistent. But economists still argue that markets can prevent unsustainable depletion through the price mechanism. Unfortunately, by the time prices rise sufficiently to impel substitution, the renewable resource may be depleted beyond recovery.

42. Thomas Robert Malthus, *An Essay on the Principle of Population; or, A View of Its Past and Present Effect on Human Happiness; with an Inquiry into Our Prospects Respecting the Future Removal or Mitigation of the Evils which It Occasions* (London: Printed for J. Johnson, by T. Bensley, 1803; new edition, very much enlarged).

43. The 1997-98 El Niño weather phenomenon illustrates this proposition rather nicely: In the ocean, many species can follow the food from their normal feeding groups; for example, blue whales, tuna, and marlin being seen and caught off the coast of northern California. Anchovy, however, are dying in great numbers off the coast of Peru, being less able to "go with the flows."

44. Dennis Meadows, et al., *Limits to Growth* (Cambridge, Mass.: MIT Press, 1972). See also Donella H. Meadows, Dennis L. Meadows and Jorgen Randers, *Beyond the Limits: Confronting Global Collapse, Envisioning a Sustainable Future* (Post Mills, Vt.: Chelsea Green, 1992).

45. Jean Dreze and Amartya Sen, *Hunger and Public Action* (New York: Oxford University Press, 1989); Amartya Sen, "Population: Delusion and Reality," *New York Review of Books* (September 22, 1994): 62–71.

46. Again, this proposition is illustrated by the 1997 famine in North Korea; to open up the country to food and markets would be to give up the raison d'être for the North Korean state.

47. But see Alexander Wendt, "Anarchy Is What States Make of It: The Social Construction of Power Politics," *International Organization* 46, no. 2 (Spring 1992): 391–425.

48. Robert Gilpin, "Economic Interdependence and National Security in Historical Perspective," in Klaus Knorr and Frank Trager, eds., *Economic Issues and National Security* (Lawrence: Regents Press of Kansas, 1977): 19–66.

49. Fred Block, *The Origins of International Economic Disorder—A Study of United States International Monetary Policy from World War II to the Present* (Berkeley: University of California Press, 1977).

50. It was called the "Grand Area"; see Laurence H. Shoup and William Minter, *Imperial Brain Trust: The Council on Foreign Relations and United States Foreign Policy* (New York: Monthly Review Press, 1977).

51. Ronnie D. Lipschutz, *What Did You Do in the Cold War, Daddy? Reading U.S. Foreign Policy in Contemporary Film and Fiction*, draft manuscript, 1996, chapter 3.

52. For a discussion of the "natural community," see Deborah A. Stone, *Policy Paradox and Political Reason* (New York: HarperCollins, 1988).

53. The distinction was never as great as claimed. The isolationists wanted to keep pernicious influences out; the internationalists wanted to keep them contained. Both aimed to avoid "contamination."

54. John G. Ruggie, "International Regimes, Transactions, and Change: Embedded Liberalism in the Postwar Economic Order," in Stephen D. Krasner, ed., *International Regimes* (Ithaca, N.Y.: Cornell University Press, 1983): 195–232. For a more recent discussion, see John G. Ruggie, "Embedded Liberalism Revisited: Institutions and Progress in International Economic Relations," in Emanuel Adler and Beverly Crawford, eds., *Progress in International Relations* (New York: Columbia University Press, 1991): 210–334.

55. Lipschutz, *When Nations Clash*, chapter 5; Beverly Crawford, *Economic Vulnerability in International Relations* (New York: Columbia University Press, 1993).

56. Enrico Augelli and Craig Murphy, *America's Quest for Supremacy and the Third World — A Gramscian Analysis* (London: Pinter, 1988); Stephen Gill, ed., *Gramsci, Historical Materialism and International Relations* (Cambridge: Cambridge University Press, 1993).

57. See, e.g., Charles P. Kindleberger, *The World in Depression, 1929–1939* (Berkeley: University of California Press, 1973); Robert Gilpin, *War and Change in World Politics* (Cambridge: Cambridge University Press, 1981); Robert O. Keohane, *After Hegemony — Cooperation and Discord in the World Political Economy* (Princeton, N.J.: Princeton University Press, 1984); Paul Kennedy, *The Rise and Fall of the Great Powers* (New York: Random House, 1988).

58. See the discussion in Joshua S. Goldstein, *Long Cycles — Prosperity and War in the Modern Age* (New Haven, Conn.: Yale University Press, 1988).

59. Susan Strange, "*Cave! hic dragones*: A Critique of Regime Analysis," in Stephen D. Krasner, ed., *International Regimes* (Ithaca, N.Y.: Cornell University Press, 1983): 337–54.

60. Robert O. Keohane and Joseph S. Nye, *Power and Interdependence* (New York: HarperCollins, 1977): 8.

61. Lipschutz, *When Nations Clash*, 129. Nor did they recognize that, given the nature of oil markets, even control of oil would not have prevented a generalized increase in prices; see Lipschutz, "Strategic Insecurity."

62. Wendt, "Anarchy Is What States Make of It"; Jonathan Mercer, "Anarchy and Identity," *International Organization* 49, no. 2 (Spring 1995):229–52.

63. Edward Said, *Orientalism* (New York: Viking, 1979): 54, quoted in Dalby, *Creating the Second Cold War*, 20.

64. Keohane and Nye, *Power and Interdependence*, 7; my emphasis.

65. Nancy Lee Peluso, *Rich Forests, Poor People — Resource Control and Resistance in Java* (Berkeley: University of California Press, 1992); Nancy Lee Peluso, "Coercing Conservation: The Politics of State Resource Control," in Lipschutz and Conca, *The State and Social Power*, 46–70.

66. World Commission on Environment and Development, *Our Common Future* (Oxford: Oxford University Press, 1987): 4.

67. WCED, *Our Common Future*, 38.

68. The continued resistance of the United States to actually controlling its emissions of greenhouse gases is testimony to the faintness of this particular hope.

69. WCED, *Our Common Future*, 8.

70. WCED, *Our Common Future*, 44.

71. Gleick, "Water, War, and Peace."

72. Gleick, "Water, War, and Peace."

73. Karen T. Litfin, chapter 1, this collection.

II

Reconfiguring Sovereignty: Case Studies

6

Forms of Discourse, Norms of Sovereignty: Interests, Science, and Morality in the Regulation of Whaling

Ronald B. Mitchell

Introduction[1]

Many environmentalists have argued that existing norms of state sovereignty promote environmental degradation. In particular, the overuse of global common-pool resources is often attributed to international legal norms that define various international commons as open-access resources and allow governments to make independent decisions about their use. To reduce such overuse, governments increasingly have sought to use international legal conventions to redefine the rights of states in areas of common jurisdiction. By redefining these rights, states are redefining sovereignty. Given the weakness of international law within an anarchic international system, however, altering legal definitions of sovereignty need not alter the actual practice of sovereignty. The question then arises: Under what conditions will de jure redefinitions of sovereignty alter the de facto practices of sovereignty that destroy the environment?

The success of efforts to alter sovereign practice by redefining sovereign rights depends upon the form of discourse used to justify the redefinition. Put differently, rhetorical justifications influence the practical legitimacy accorded to a nominally binding international legal norm of sovereignty. The case of international regulation of whaling provides one illustration that a redefinition of sovereignty established through a discourse involving scientific, causal arguments alters sovereignty as practiced more readily than the same redefinition established through interest-based argument, which, in turn, alters sovereign practice more readily than such a redefinition established through moral or principled arguments.[2] Causal belief–based discourse leads states to accept and abide by new sovereignty norms if those states share causal beliefs and an acceptance of the process by which they are validated and changed. Interest-based discourse leads states to accept and abide by new sovereignty norms if those states share an acceptance of the structure of power and interests in the international system. Principled belief–based discourse leads states to accept and abide by new sovereignty norms if those states share an acceptance of the values and principles underlying that discourse.[3] In the whaling case, causal, scientific discourse appears to have fostered the norm of collective decision making most. The whaling case provides a useful laboratory for analyzing these issues because the three discourses are temporally separated. In other environmental issues, the three discourses tend to be more intertwined. Thus, whether findings from the whaling case generalize to other environmental issues remains an open question for future research.

During different phases in the whaling regime's history, member states have used scientific, interest-based, and moral arguments to justify a single redefinition of sovereignty. At the inception of the International Whaling Commission (IWC) in 1946, the whaling states negotiated a de jure change to international legal norms of sovereignty: States agreed to determine the quantity, type, location, timing, and methods of taking from the international common-pool resource of whales through collective rather than independent decision making.[4] The IWC has had mixed success in its subsequent efforts to induce corresponding de facto changes in sovereignty. Initially, those advocating collective decision making relied almost exclusively on interest-based arguments. States submitted to this new norm of sovereignty to a limited extent but reverted to independent decision making whenever their short-term interests seemed at risk.

During a second phase, scientific arguments involving causal ideas about the state of whale populations and the likely consequences of maintaining then-current levels of whaling gained influence. These scientific arguments led states, despite their reluctance, to increasingly allow collective IWC decisions to constrain their whale hunts, even though neither their short-term interests nor the industry's "tragedy of the commons" dynamics had changed. During a third phase, morally based arguments for a complete discontinuance of whaling came to dominate IWC debates. Initially, when the policy recommendations of moral arguments coincided with those derived from scientific arguments, morally based arguments appeared to reinforce the commitment to practicing the norms of sovereignty embodied in the treaty. However, in the current fourth phase, divergence between the policy recommendations of moral and scientific discourses have shown that morally based arguments fail, when operating alone, to induce governments to put agreed-upon legal norms of sovereignty into practice. Indeed, they induce a reactive resistance that has led states to explicitly reject the legitimacy of the norm both by word and by deed.

Norms of Sovereignty

We need not engage the recent debate regarding whether environmental issues are eroding state sovereignty, to recognize that environmental treaties, at a minimum, redefine and reconceptualize sovereignty.[5] Norms of sovereignty can be defined as the set of standards governing a state's legitimate rights and authority within its borders, within the borders of other states, and in international areas outside any state's borders.[6] States use treaty negotiations as one process, inter alia, by which to reconstruct the intersubjective consensus regarding these norms and "the actual content of sovereignty, the scope of the authority that states can exercise."[7] Nations especially have taken to using international law to define and redefine what states can legitimately do and not do in the world's "not yet sovereignized" commons—such as the oceans, the atmosphere, and Antarctica.[8] The standards that states have agreed, through treaties, should guide behavior constitute de jure norms that can be distinguished from the de facto norms or standards that actually guide behavior.[9]

The Four Phases of Discourse in the IWC[10]

Traditional legal norms of sovereignty hold that states can take ocean resources in international areas outside territorial limits "under a doctrine of freedom of access to them (a freedom which can be limited only with the consent of the participant state)."[11] In 1946, fifteen nations negotiated the International Convention for the Regulation of Whaling (ICRW) in an effort to avoid repeating the overexploitation of whale stocks that had preceded World War II.[12] The convention formed an International Whaling Commission (IWC) of all member states to develop an annual "schedule" of restrictions on the quantity, type, and methods of whale catches. By so doing, the ICRW established a new de jure norm that delegitimized the then-current practice of each nation independently deciding the size and manner of its whale catch. Nominally, IWC members merely negotiated the rules governing various parameters of each year's whaling. Yet, they thereby have been engaged in the metaprocess of inducing whaling states to engage in, and accept the outcomes of, these negotiations. This larger enterprise involved an effort to legitimize a new de jure norm of sovereignty, transforming it into a de facto practice, of states submitting to an ongoing process of collective, rather than independent, decision making about the limits to place on access to whale stocks.

The IWC has evolved through four different phases. From 1946 until the late 1960s, regulatory limits were established based on the preferences and power of what was essentially a "whalers' club."[13] Scientific arguments exercised little influence in IWC debates, and morally based environmental arguments were completely absent. Nations negotiated collective quotas and made their fleets comply with those quotas only when they believed doing so furthered their short-term economic interests.[14] A second phase began in the late 1960s, as increasing scientific expertise and consensus on whale population dynamics produced quota recommendations that diverged from those dictated by interest-based bargaining between competing economic interests. Nations increasingly accepted these alternative quotas derived from a scientific discourse, reverting to independent decision making with decreasing frequency. The IWC's third phase was initiated as environmental NGOs introduced a new, morally based discourse. This discourse progressively gained influence and, by 1982, produced a moratorium on commercial whaling that had little scientific rationale. At first, whaling states reluctantly accepted the moratorium, refraining from commercial whaling. More recently, however, a fourth phase has emerged in which exclusive reliance on arguments grounded in moral beliefs have decreased the whaling states' commitment to the process of collective decision making, with scientific and commercial whaling outside IWC purview increasing in frequency.

During all four phases, the IWC sought to supplant a norm of independent decision making with one of collective decision making. The success of these efforts varied across the phases, correlating with shifts in the discourse used to justify this new practice of sovereignty. The next four sections of this chapter analyze each phase by reference to three common questions. First, did the IWC produce meaningful, collectively determined quotas, i.e., did IWC quotas diverge from what would be predicted from simply aggregating the independent decisions of member states? Second, what discursive rationales were used to convince states to accept IWC quotas and, in so doing, to accept a reduction to the traditional scope of sovereign decision-making power? Finally, did states actually accept these quotas?

The Dominance of Instrumental Discourse

Traditional international norms, treating whales on the high seas as a nonexcludable common-pool resource, had created the familiar incentives and classic problems of a tragedy of the commons. By 1946, a whaling industry increasingly feeling the costs of its own overexploitation of whale stocks sought "mutual restraint, mutually agreed upon."[15] To overcome these problems, the IWC began setting annual global quotas on the number of whales that could be taken. Member states, however, did not give the IWC the power to allocate this global quota among them. This management approach encouraged overinvestment in whaling equipment as each firm tried "to take as many whales as it could before the season ended." This overinvestment, in turn, created a self-perpetuating dynamic that made it "hard to persuade managers that lower quotas were needed."[16]

From its inception until the late 1960s, the IWC's quotas diverged little, if at all, from the catches one might have expected in their absence. The initial quota of 16,000 blue whale units (BWUs) (based on an arbitrary scientific "guesstimate") was one-third lower than catches just prior to the hiatus in whaling during World War II.[17] But this figure still exceeded then-current capacity (due to wartime losses) and did not pose a meaningful economic constraint.[18] Whaling interests did not even argue for a higher figure because they did not expect to catch the full quota.[19] Until 1962, quotas never went below 14,500 BWU, and thereafter they continued to be "too high . . . at the insistence of the whaling countries, reflecting the demands of their whaling companies, despite the obvious decline in the whale stocks."[20] Until 1965, the close correspondence between each year's quota and the previous year's catch data suggests that quotas were simply best estimates of what could be caught economically rather than genuine attempts to overcome collective action problems or to respond to scientific warnings.[21] The declining quotas of this phase appear to have been driven largely by "the fact that the whaling nations were no longer able to fulfill their quotas."[22] Although whaling states recognized that global quotas encouraged overcapitalization, whaling interests frustrated attempts to negotiate national quotas until "most Antarctic whaling nations no longer found it profitable to continue their operations."[23] In essence, little collective decision making was actually occurring.

When collective decisions were made about quotas, they were founded on arguments about economic power and interests, rather than science. For example, despite strong evidence of declining humpback stocks, the IWC's first meeting in 1949 removed the original 1946 schedule's ban on taking humpback.[24] By 1950, several member states viewed the 16,000-BWU quota as "too high," but these voices were overridden.[25] Although whaling industry pressure could produce quotas that reflected their economic interests, it could not thereby resolve the underlying collective action problem. IWC membership was dominated by whaling states, and the available scientific advice was both highly uncertain and industry-based.[26] Not until 1961 did the IWC seek advice on population dynamics from an expert panel that was independent of the whaling industry. After fifteen years, IWC members finally began reducing quotas, going from 15,000 BWU in 1962 to 3,500 by 1966.[27] However, even these seemingly deep cuts rejected scientific recommendations for far deeper cuts and for replacing the BWU with species-specific quotas; scientists saw their work as almost "entirely ignored."[28] Economic interests continued to dictate quotas, which remained high "not because governments were unaware of the cetologists' concerns but, rather, because there was no other way to sustain the interwhaling state agreement."[29]

Even though the regime posed relatively minor constraints on their sovereignty, states consistently followed the traditional norm of independent decision making whenever following the norm of collective decision making would have contradicted short-term economic interests. Whaling interests regularly threatened, or carried out threats, to ignore collective decisions by opting out of specific rules or withdrawing from the IWC. A 1954 ban on the taking of blue whales in the North Pacific and North Atlantic was rendered meaningless when every nation actually hunting blue whales in those areas opted out from the prohibition.[30] Norway and the Netherlands withdrew from the IWC from 1959 until 1962 to protest proposed quotas they viewed as overly restrictive.[31] Other whaling states remained outside the regime. Panama, Chile, and Peru all found ways of avoiding IWC quotas.

In short, during this initial phase, states failed to establish catch levels different from those that would have occurred without the IWC, produced decisions through traditional interstate bargaining determined by economic power and interests, and ignored collective decisions whenever it suited their short-term interests. IWC members negotiated quotas in terms of economic self-interest, which failed to produce a de facto practice corresponding to the de jure norm of collective decision making enshrined in the ICRW.

The Dominance of Causal Discourse

From the 1960s through the 1970s, industry catch data increasingly confirmed scientific warnings that lower quotas were vital to maintaining whale stocks that could support commercial whaling. No longer could industry "discredit the cetologists' case for quota reductions on scientific grounds."[32] Initially, this led to somewhat different quotas, but little difference in behavior. IWC members consistently rejected the Scientific Committees recommendations to abandon the BWU system, but they did adopt scientifically recommended limits on specific species and regional stocks. In 1963, IWC members rejected the Scientific Committees recommendation to ban taking of blue whales in the Antarctic but adopted a ban on taking humpbacks. A year later, the Antarctic blue whale ban was adopted, though all the countries whaling in that region subsequently filed objections.[33] Evidence that these species could still be economically harvested and that adoption of these bans did not merely codify existing interests comes from the fact, known at the time, that non-IWC states (Chile and Peru) were taking blue and humpback whales and the recently revealed fact that Soviet whalers were catching large numbers of humpbacks during these years.[34]

In 1967, the IWC set the first quota below scientific recommendations.[35] A New Management Procedure (NMP) using scientific estimates of sustainable yields to set quotas was regularly discussed, eventually adopted in 1974, and implemented in 1978.[36] Thus, between 1960 and 1980, scientific arguments gained power, producing a transition from quotas as mere aggregations of the expected individual catches of member states to low overall and species-specific quotas, which clearly resulted from collective, interdependent decision making. Causal discourse's increased influence arose from scientists' growing consensus regarding the decline in whale stocks, the whaling states' growing recognition that economically dictated quotas were failing to reduce overexploitation, and the nonwhaling states' growing willingness to reject the interest-based discourse of whaling states.[37] Scientific evidence transformed quota debates from revolving around, How will whaling states respond to an excessively low quota?, to revolving around, How will whale stocks respond to an excessively high quota? Setting quotas in response to the former required only diplomatic skill. Setting quotas in response to the latter required scientific advice. It also, however, involved a shift from short-term economic reasoning to longer-term ecological reasoning. Adoption of the NMP with its species-specific quotas reflected a new willingness of IWC member states "to give science [and the Scientific Committee's recommendations] a much more prominent, though not exclusive, place in the decision-making system."[38]

By the 1970s, states successfully used scientific arguments to resist economic pressures for higher quotas as well as growing environmental pressures for a blanket moratorium. Surprisingly, whaling states joined other states at the 1972 United Nations Conference on the Human Environment in unanimously recommending a ten-year moratorium on all commercial whaling. In the ten years following that recommendation, moratorium proposals regularly came before the IWC. However, the commission's Scientific Committee consistently rejected such proposals because a moratorium on all whales would "directly conflict with these new [NMP] principles" of using scientific knowledge to manage individual whale stocks.[39] The IWC's second phase exhibited a power-based bargaining dynamic quite different from that of the first phase. Whaling states no longer merely withdrew or opted out when quotas fell below their preferred outcomes. Nor did American pressure lead to adoption of a blanket moratorium. Science provided the focal point for compromises that kept whaling states "at the table" while pushing the moratorium off the IWC agenda until 1979.

IWC members not only adopted meaningful limits but increasingly conformed their behavior to those limits. The de jure norm of collective decision making was finally, albeit slowly, becoming a de facto norm. Whaling states' objections to global quotas ceased by the mid–1960s and objections to species-specific quotas ceased, with one exception, after the late 1960s.[40] By 1974, the commission was limiting catches of every species of whale without a single state protesting.[41] Even when member states lodged objections, their behavior increasingly exhibited restraint. Japan, the Netherlands, and Norway caught few, if any, blue or humpback whales after objecting to regional bans on these species in the late 1960s. Although the Japanese and Soviets objected to a regional quota of 5,000 minke whales, they together took only 7,700 whales.[42] Japanese whalers caught only two sperm whales before retracting their objection to a 1981 ban in response to U.S. pressure.[43] Large catches by Chile, Peru, and the Soviet Union confirm that Japanese, Dutch, and Norwegian whalers were actually practicing restraint, rather than merely failing to find whales.[44] Indeed, the failure of some states to respect IWC quotas demonstrated the economic viability of, and exacerbated the incentives for, continued whaling by all whaling states, since restraint was clearly not being reciprocated and species might well go extinct anyway.

The norm of collective decision making broadened as well as strengthened during this period. As even small-scale operations could threaten specific species stocks, IWC states increasingly sought to bring the whaling of nonmember states (e.g., Brazil, Chile, Panama, Peru, South Korea, and Spain) under IWC jurisdiction. Though nonmember whaling was clearly legal under international law and conformed with traditional norms of sovereignty, such free-riding undercut the developing norm of collective decision making regarding whaling. IWC members unanimously banned the import of whale products from, and the transfer of whaling ships and equipment to, nonmember states.[45] At the same time, the United States used threats of economic sanctions to induce Chile, Korea, Peru, and Spain to join the IWC.[46]

During this phase, collective decisions increasingly ran counter to members' short-term economic interests. Although perhaps too late for some species, IWC states had largely achieved, in law and in practice, the "mutual restraint mutually agreed upon" essential to overcoming the tragedy of the commons.[47] The shift arose from an increasing willingness of commission members to accept causal scientific, rather than self-interested economic, discourse as an appropriate basis for collective decisions. More whaling states became IWC members, and members objected to commission decisions far less frequently. If the IWC's initial history had been characterized by material interests determining both state positions and state action, with science opportunistically brought to bear to serve those predetermined positions, this second phase was characterized by positions and actions being developed in response to recommendations produced by new scientific paradigms, arguments, models, and data. Quotas and catches increasingly came to reflect calculations of population sustainability rather than calculations of economic viability. Scientific discourse, which initially had served merely as a new source of justification for old interest-based positions, transformed those positions to reflect scientific as well as material factors. By the end of this period, whaling states had come to accept collective decision making as a de jure and a de facto norm, not in response to political pressures but increasingly in response to ecological ones.

Before accepting such an analysis, however, two alternative explanations should be explicitly examined. American sanctions authorized under the Pelly and Packwood-Magnuson amendments and supported by domestic antiwhaling sentiments have been regularly turned to in the 1990s and might suggest that such economic pressure rather than scientific arguments explain increasing whaling state compliance with IWC decisions. However, the empirical evidence refutes such a claim. IWC whaling states were accepting and conforming to steeply declining commission quotas before the United States had passed such legislation, let alone invoked it. Antarctic whaling quotas *and catches* had already declined from 15,000 BWU in 1963 to 1,475 BWU by 1974,[48] the first year in which the United States threatened economic sanctions in support of IWC regulations. That threat against Japan and the Soviet Union was never carried out and the United States did not threaten similar sanctions again until the late 1978 sanctions to induce Chile, Peru, Korea, and Spain to become IWC members.[49] It might also be claimed that scientific arguments merely helped clarify these states' material interests, i.e., science clarified states' interests without actually altering how they were defined. Such an argument runs counter to both the logic of the tragedy of the commons and the IWC experience. Discussions of population levels and maximum sustainable yields were not simply brought to bear in support of economic interests developed independently. Indeed, whaling states' interests in negotiating higher quotas or violating lower ones remained strong or increased during this phase: Since very low and declining whale stocks made extinction possible even without further whaling, and unregulated and clandestine whaling could still occur, each whaling state had strong incentives to hunt the last of the whales before they became extinct. Although no other country appears to have acted on these incentives, recent revelations of Soviet violations confirm that these incentives remained strong. The strategic structure of the tragedy of the commons had, if anything, been exacerbated by the declining resources on the commons. Scientific information did not merely clarify the "payoffs" of the game by improving estimates of expected catches, but instead altered states' understanding of the structure of the game itself. Scientific perspectives altered the discourse of states within the IWC from one concerned with the fundamentally political question of the short-term response of other states to one concerned with the fundamentally scientific question of the long-term response of the whale stocks.

The Dominance of Principled Discourse

Adoption of a commercial moratorium in 1982 illustrated the growing impact of a moral discourse that had been part of IWC debates since the 1970s. Most whaling states considered any moratorium economically undesirable and most cetologists considered a commercial moratorium that failed to distinguish between species scientifically unsound. However, a temporary moratorium appealed to those cetologists who sought time to improve the accuracy and certainty of their models; conservationists who sought to ensure the sustainability of whale stocks; and preservationists, deep ecologists, and animal rights activists who sought to protect all whales from human predation.[50] These actors' agreement on the desirability of a temporary "zero quota" on commercial whaling masked their fundamental disagreement over the principles that should guide the IWC's collective decision making.

By the late 1970s, collective IWC decision making took on new meaning. The whale had become "the most poignant symbol of the world environmental movement."[51] Recognizing that the ICRW did not restrict membership and that a three-quarters majority could amend the schedule, environmental groups made a top priority of convincing nonwhaling states to join the IWC and vote for the zero quotas.[52] Simultaneously, the United States was using economic and diplomatic pressure to induce nonmember whaling states to become commission members so that regime rules would apply more broadly. Together, these efforts transformed IWC membership from eight nonwhaling and eleven whaling members in 1978 to twenty-seven nonwhaling and twelve whaling members by 1982.[53] Collective decisions now needed to reflect the interests of states with neither a current nor an historical interest in whaling. Members no longer shared a common goal for the regime: Older members supported the traditional goal of conserving whale species to preserve the whaling industry, while many new members wanted to preserve all whales and end the whaling industry.

The commercial moratorium constituted a major rejection of both whaling interests and scientific guidance. Although much of the debate on the moratorium was framed in scientific language, support for the moratorium was grounded in moral, rather than scientific, principles. For example, although the moratorium provided for regular scientific reviews and reestablishment of catch limits based on a comprehensive scientific assessment no later than 1990, the IWC's own Scientific Committee and the Food and Agriculture Organization criticized the moratorium as deriving from aesthetic and moral principles with "no scientific justification."[54] Most scientists viewed a temporary moratorium as essential to the recovery of some species. But a *commercial* moratorium on *all* species lacked a sound scientific basis, simultaneously overprotecting some unthreatened species while underprotecting others that were threatened by even small-scale noncommercial aboriginal whaling. The moratorium clearly constituted a different logical foundation for IWC decisions than either the independent interests that drove the first phase or the scientific advice that drove the second.

The influence of moral discourse on IWC debates does not imply that they had influence on the positions of all commission members. Indeed, Spain was the only whaling state that voted for the moratorium. Japan, Norway, Peru, and the Soviet Union immediately filed objections to it, basing their arguments in part on the requirement that "amendments be based on scientific findings."[55] Although domestic political forces ensured that the environmental principle of "saving the whales" informed the American position on the moratorium, the United States used far more material resources to induce whaling states to accept the decision of the IWC majority. Economic threats of reducing imports from or fishing rights of whaling states led Peru to remove its objections immediately and Japan to follow suit in 1987. Norway and the Soviet Union maintained their objections but stopped commercial whaling by 1987 and 1988, respectively.

The ascendance of moral argument reflected two trends. Scientific recommendations had become increasingly uncertain by the late 1970s because the New Management Procedure required data in excess of what member states were willing to provide.[56] "Internal disagreements about which model to use and how to interpret the data made it difficult to give the unified advice necessary to counter the influence of either the industry-oriented members of the IWC Scientific Committee or the environmentalists."[57] This increasing uncertainty and decreasing consensus were united with a shift toward preservationist, from conservationist, values even among IWC scientists, and increasingly close ties between many scientists and environmental groups.[58] In addition, activists committed to the rights of whales, who had previously relied on the IWC's dominant scientific rhetoric, increasingly shifted to an unabashedly moral discourse that saw whales as so unique that "they should not be killed at all [and] . . . had rights, comparable to human rights, to exist in the oceans without being exploited in any way whatsoever."[59] Setting whale quotas was no longer a scientific question but an "ethical question and whales should not be killed because it is unethical to kill them."[60] The "totemization" of whales transformed nominally scientific debates within the IWC into polarized political and moral debates.[61]

Surprisingly, despite their objections to a moratorium that reflected neither their economic interests nor scientific reasoning, the world's whaling fleets largely conformed with IWC decisions during the 1980s. Rather than following the 1950s' pattern of rejecting the IWC process and the moratorium outright, all whaling states chose to constrain their whaling within the limited legal outlets of scientific and aboriginal whaling. Not a single state withdrew from the IWC for ten years and U.S. pressure forced these states to end their commercial whaling, even if not to retract their objections to the moratorium. Although whaling states had issued relatively few scientific permits before the mid-1970s, Iceland, Japan, Korea, and Norway submitted scientific permit proposals to the Scientific Committee regularly in the late 1980s and early 1990s. Although these states issued these permits even after the committee rejected the proposals, the permits constituted quite limited takes (between 20 and 350 whales per year) compared to what might have been expected in the moratorium's absence.[62] Although extensive use of scientific permits constituted a clear rejection of the moral "spirit" of the moratorium, all the whaling states except Iceland limited even their scientific whaling to the minke whale stocks which most cetologists considered quite healthy. Similarly, whaling states increasingly drew attention to the seemingly arbitrary distinction between "aboriginal" whaling allowed under the convention and "small-type coastal whaling" conducted by Japanese and Norwegian communities. Thus, during the 1980s, the whaling states proved unwilling to completely reject (by withdrawing) a collective decision for a moratorium that directly contradicted their interests while seeking ways to continue some whaling within the IWC framework.

In this third phase, the collective adoption of a commercial moratorium clearly diverged from what whaling states would have decided independently as well as from scientific guidance identifying the need for discriminating, species-specific quotas. The decision to adopt a moratorium instead was based on a moral and ethical discourse. Moral arguments from home or abroad to "save the whales" could influence the positions of those states that lacked countervailing economic interests and this, when coupled with the IWC's three-quarter–majority voting rule, produced a collective decision based in moral rather than economic or scientific interests. Given earlier experience, one might have

why

expected whaling states to simply reject the commission process and the moratorium outright. Yet whaling states took pains to stay within the letter, if not the spirit, of the limits dictated by the regime. Unlike the second phase, during which scientific discourse seemed to cause the behavioral conformance with the regime, the behavioral conformance of this phase seems likely to have stemmed from factors other than the rationale and discourse on which the collective decision was founded. Whaling states remained largely uncompelled, and indeed antagonized, by the moral arguments raised by environmentalists. Their behavior appears to have conformed with IWC dictates for three reasons: increasing U.S. economic and political pressure brought to bear to support the moratorium; a, perhaps inertial, "regime-mindedness" of whaling states valuing the long-term benefits of collective IWC decision making based on scientific discourse to solve the common-pool resource problem despite decisions that contradicted short-term interests; and a limited "room to maneuver" provided in the interim through scientific permits. Thus, during this transitional phase, principled discourse appeared to dictate the decisions of the IWC without contributing significantly to the behavioral conformance of whaling states with those decisions.

The Rejection of Principled Discourse

Since 1990, the IWC has entered a new phase in which whaling states appear increasingly willing to accept only those decisions grounded in scientific, rather than moral, discourse. When the IWC adopted the moratorium on whaling, scientific uncertainty about whale stocks and population modeling had created a debate that pitted the economic interests of whalers against the moral interests of various environmental groups. Scientific dissensus forced all sides to resort to arguments that the other side could interpret simply as interest-based. By the early 1990s, however, a new and strong scientific consensus emerged that certain species of whales could be hunted at limited levels without threatening survival of those species. This new consensus has brought the fundamental conflict in the principles of whalers, conservationists, and preservationists to the fore. The moratorium had held this conflict at bay by temporarily halting the practice of whaling without rejecting the principle of treating whales as a resource. As a policy that could be "all things to all people," the temporary halting of whaling could be viewed as a means of allowing stocks to recover to optimal yields for harvesting, a means of allowing stocks to recover from the brink of extinction, and an end in its own right. The moratorium's requirement for future scientific review established causal beliefs as the legitimate discourse for collective decisions even while the moratorium itself overrode then-current scientific recommendations for more discriminating, stock-by-stock quotas. The initial moratorium decision constituted a classic example in which "the cultural role of science as a key source of legitimation means that political debates are framed in scientific terms; questions of value become reframed as questions of fact."[63]

In the second phase, whaling states had submitted to IWC decisions that they opposed because they were consistent with scientific advice. The adoption and maintenance of a moratorium based on a moral discourse, and contradicting scientific advice, made these states increasingly skeptical that future IWC policy would be based on causal beliefs. That skepticism faced its first test with evaluation of the moratorium's effect in 1990.[64] By that year, many cetologists contended that minke whale stocks could sustain limited commercial harvests. Nevertheless, the IWC rejected Norway's proposal to recommence commercial whaling within scientifically prescribed limits.[65] By 1991, Scientific Committee estimates confirmed the strength of minke stocks.[66] However, the moral discourse of the 1980s had shifted the focus of commission debate from whether whale stocks *could* sustain whaling to whether they *should* sustain whaling. In 1992, the IWC adopted an improved Revised Management Procedure (RMP) that promised to provide more accurate quotas making it "possible to authorize a catch that year."[67] IWC members reiterated their support for the RMP in 1993 and the commission's secretary-general stated that, "In all reasonableness, we would have to say that a commercial catch could be taken without endangering [minke] stocks."[68] Despite these assessments, the IWC has extended the moratorium in every year since 1990. The commission has refused to authorize Japanese and Norwegian coastal whaling and, in 1994, adopted a whale "sanctuary" that outlawed both scientific and commercial whaling in Antarctic seas.[69] Maintenance of the moratorium, adoption of the RMP, and creation of the sanctuary demonstrate the IWC's continuing ability to arrive at collective decisions, despite increasingly vehement opposition from whaling states. This ability owes much to the IWC's three-quarter–majority rule, which made adoption of the moratorium difficult but its repeal even more so.

The growing scientific consensus that limited whaling would not threaten certain species removed the previously plausible argument that continuing a blanket moratorium was warranted by scientific uncertainty. During the IWC's second phase, all sides had appealed to the same scientific principles, even while arguing over the scientific "facts." Now, however, different positions explicitly reflect different underlying principles. Environmental groups began to reject the scientific discourse they had previously embraced: "Even if humanity thinks that it has an ironclad 'scientific' banner under which to kill the whales, is that enough? Is the paradigm under . . . which it is okay to take the maximum number of a particular species according to a complicated calculation of 'sustainability,' defensible?"[70] In contrast, Japan's IWC representative sought to reject such a basis for commission decisions and revert to scientific principles: "We believe science and we believe scientists. We should not permit religious arguments in this field."[71] Norway's representative claimed IWC decisions now reflect "cultural imperialism imposed by the majority of the members of the IWC on the local communities of the nations and peoples who want to exercise their sovereign cultural right to be different."[72] IWC decisions have increasingly reflected moral principles, even explicitly rejecting scientific rationales at times. For example, the Antarctic sanctuary provision explicitly required that scientific advice be ignored, applying "irrespective of the conservation status of baleen and toothed whale stocks in this sanctuary."[73] Philip Hammond resigned as chair of the Scientific Committee because of the consistent rejection of the committee's advice.[74] The transitional third phase during which IWC decisions reflected a coincidence of interests among those concerned about scientific uncertainty, stock recovery, and moral principles, became a fourth phase in which, increasingly, only the last of these was relied on.

The overarching de jure norm of collective decision making remains intact in the ICRW, but the actual practice as exhibited by whaling state behavior, i.e., the de facto norm, has begun to erode. The reluctant willingness of whaling states to conform to IWC dictates during the 1980s began to unravel after 1990. Iceland withdrew its membership in the IWC in 1992, and Japan and Norway have threatened to do so because of the commission's rejection of scientific arguments.[75] Norway, having maintained its objection to the moratorium, recommenced commercial whaling the day after the 1992 decision to extend the moratorium.[76] Notably, even in reverting to independent decision making, Norway recognized the legitimacy of scientific discourse, restricting its hunt to minke whales, setting a catch limit below that recommended by the Scientific Committee's Revised Management Procedure, and gaining Hammond's scientific approval before proceeding. Frustrated by the "feeling that a number of the IWC members over several years had not been negotiating in good faith," Norway, Iceland, Greenland, and the Faroe Islands established the North Atlantic Marine Mammal Commission (NAMMCO).[77] NAMMCO adopted the RMP as a guide for its decision making, embracing an IWC-approved scientific model the recommendations of which the commission itself has rejected in the face of moral arguments.[78] Japan is considering establishing a Pacific Ocean counterpart to NAMMCO.[79] Even subsistence Inuit whalers view NAMMCO as a desirable complement, perhaps alternative, to the IWC."[80] By replacing a global forum with a regional whale management regime, this move reaffirmed these states' commitment to a sovereignty norm of making whale management decisions collectively while rejecting that such collective decisions can or should be based on moral principles. Although, to date, only Norway has recommenced whaling, the elements that held whaling states in check during the 1980s appear to be weakening.

Whereas the late 1950s had seen Norway and the Netherlands operate outside the IWC in response to decisions reflecting excessive scientific influence, Norway, Iceland, Greenland, and the Faroe Islands have now chosen to operate outside the commission in response to decisions reflecting insufficient scientific influence. Growing scientific certainty has revealed a deep-seated divergence in principled ideas.[81] Removing the scientific support for a moratorium converted a nominally scientific debate into a fundamentally moral and principled one.[82] But whaling states appear less willing to abide by collective decisions based on moral principles that they do not accept. Indeed, the actions of whaling states in the last five years demonstrate that they accept the need for collective decisions, but only when those decisions reflect scientific discourse.

Certainly in the shorter term, the legitimacy of collective decisions within the IWC, and the willingness of states to accept the corresponding constraints on their sovereignty, "rests on expert scientific resource management, not beliefs held by members of the IWC about the sanctity of whales."[83]

Analysis

How do arguments based in material interests, scientific knowledge, and moral principles explain the outcomes we observe in international regulation of whaling? Specifically, how did these three different forms of discourse influence the willingness of states to accept, in practice, a redefinition that placed new limits on traditional norms of sovereignty? This section analyzes the whaling experience and less systematic evidence from other cases to develop four propositions regarding the conditions under which different discourse types are likely to lead states to practice new norms of sovereignty.

The International Convention for the Regulation of Whaling was initially established to resolve a classic tragedy of the commons involving overfishing of whale stocks. To achieve Hardin's "mutual restraint, mutually agreed upon," states had to forego their sovereign right to set national whaling policies independently. The traditional sovereignty norm of free access to high seas resources needed to be replaced with collectively decided constraints on access. For the two decades after the IWC's formation in 1946, the rhetoric supporting acceptance of this new norm appealed to the economic interests of whaling states. In classic collective action fashion, quota levels reflected a competition between fear that maintaining high catch levels would decimate whale stocks and destroy the industry in the long term, and fears that reducing catch levels would impose even greater immediate costs on an overcapitalized industry. These interest-based arguments failed to compel states to conform their behavior to a new norm of collective decision making, at least whenever doing so conflicted with their immediate economic interests. Those states powerful enough to do so regularly reverted to the traditional norm of independent decision making. The regular withdrawal or opting out of collective decisions during the early years of the IWC document the inability of interest-based arguments to keep governments at the table, let alone influence their behavior. In short, the discourse of IWC diplomacy was essentially epiphenomenal, with real bargains being struck only when they accurately reflected the outcomes that power and interests would have dictated anyway. Other fisheries regimes have experienced similar phases during some, if not all, of their evolution. We can also observe similar influences in the European acid rain debate: Weak laggard states quickly acceded to reduce sulfur dioxide emissions by 30 percent once Germany declared its willingness to do so. Germany's actions signaled these states that they were likely to be forced to accept these commitments anyway in the near future.[84]

PROPOSITION 1. *Interest-based discourse leads states to accept new norms of sovereignty that conflict with their short-term interests only if that discourse convinces them that developing patterns of power and interests will force reluctant states to accept such norms in any event.*

Beginning in the late 1960s, the rhetoric within the IWC gradually shifted from economic interests to scientific causal beliefs. Faced with the failure of quotas arrived at through bargaining among economic interests, industry and governments increasingly sought out scientists and accepted scientific discourse in identifying solutions. Although scientific recommendations were regularly ignored, growing scientific discourse progressively strengthened a commitment to collective decision making. Objections and withdrawals became less frequent. Even when scientific consensus was weak, debates revolved around scientific knowledge and uncertainty. Although IWC members often used scientific uncertainty to justify interest-based disregard for scientific advice, they began reverting to independent decision making less often. As late as the 1980s, whaling states conformed their behavior to distasteful collective decisions because scientific uncertainty made outright rejection of a moratorium untenable within the dominant rhetorical context. Faced with uncertainty and conflicting causal beliefs, whaling governments accepted science as an appropriate discourse and scientific consensus as an arbiter of policy conflict. The legitimacy states accord the scientific paradigm led them to continue negotiating rather than revert to traditional norms of independent decision making. Since 1990, whaling state actions have confirmed that their commitment to collective decision making is contingent on those decisions being based in scientific, rather than interest-based or principle-based, discourse. Scientific arguments have more readily compelled states to act collectively despite conflicts with short-term interests.

Why should scientific discourse prove more compelling than interest-based discourse? For whalers, overexploitation of whale stocks clearly constituted a traditional collective action problem. Within an interest-based discourse, analyzing the cause of declining whale catches focused states' attention on the current allocation problem, i.e., how to allocate existing stocks of whales. Framed in those terms, the dominant discursive focus became how other states would respond to one's own restraint. Thus, outcomes were seen as a function of short-term strategic interaction among states. Each state could plausibly convince itself that other states might exercise restraint, allowing it to reap the temptation payoff of defection in a prisoners' dilemma problem, a strategy most evident in the Soviet Union's clandestine catches of the 1960s. Since, in the short term, the whale stock could be considered a given, states could ignore nature's response to current overwhaling in calculating short-term payoffs. Scientific discourse, however, refocused attention on the concomitant future provision problem that plagues common-pool resources, that is, how to ensure adequate whale stocks for the future.[85] Framed in these new terms, nature's future response to current overwhaling could not be ignored. Scientific discourse also led states to recognize that nature, unlike other states, could not be deceived into providing the temptation payoff in response to their defection. Similarly, consider how scientific findings demonstrating that initial international regulations would not adequately protect the stratospheric ozone layer led to adoption of more stringent amendments, or how scientific discoveries helped overcome interest-based resistance to cleanup of the Mediterranean Sea.[86] Of course, the power of scientific discourse to influence behavior depends on the expected response of the natural system, the level of scientific consensus regarding that response, and the level of acceptance of that consensus by member nations. Scientific discourse facilitated collective decisions and conformance with those decisions in situations in which interest-based discourse did not, by transforming the calculus from one about how other states might respond to one's own overwhaling to how nature would respond.

key word

PROPOSITION 2.   *Scientific discourse leads states to accept new norms of sovereignty that conflict with their short-term interests when sufficient scientific consensus and acceptance of that scientific consensus leads states to focus their attention on how nature will respond to their actions rather than on how other states will respond.*

Beginning in the late 1970s, a growing international environmental movement reframed the debate over whaling, within and outside the IWC, in terms of principled moral beliefs. A widespread consensus on a temporary moratorium could be constructed on the basis of scientific discourse. However, a permanent, global ban on whaling could only be constructed on the basis of a moral discourse that rejected the temporary, species-specific bans supported by scientific discourse. By the 1990s, although the policy recommendations from these two discourses had diverged significantly, power, membership, and IWC voting rules prevented efforts to bring commission decisions back in line with the recommendations of scientific advisers. In response, whaling states have, become increasingly unwilling to accept IWC decisions. Recent whaling state rhetoric and behavior, especially the development of NAMMCO and its reliance on the IWC's scientific guidelines in the RMP, confirm that these states are not rejecting collective decisions per se, but rather selectively rejecting those collective decisions grounded in moral, rather than scientific, discourse.

For moral argument alone to lead a state to accept new norms of sovereignty and behavior requires that state to internalize a commitment to the underlying moral principles. Without such a commitment, "transforming state practices has come about [only] as a result of linking principled ideas to material goals."[87] Such tactics quickly become, at least to the target, identical to traditional interest-based arguments. It may be that the deep-seated transformation and internalization of values inherent to the logic of moral discourse simply takes longer to affect behavior than other forms of discourse, and that whaling states will, over time, accept the currently ascendant moral position. However, if the outcome of a negotiation grounded in interest-based discourse depends on the distribution of power, and the outcome of a negotiation grounded in causal belief–based discourse depends on scientific consensus and evidence, the outcome of a negotiation grounded in principled or moral discourse seems to look the same and depend on the same factors as those grounded in interest-based discourse. Indeed, moral discourse in the whaling case has proved even less successful than interest-based discourse because it has evoked a reactive resistance from states that do not share the ascendant moral values.[88] In response to what Iceland, Japan, and Norway increasingly perceive as ecocolonialism and cultural imperialism, cultural pride and the assertion of sovereignty appear to drive behavior even as the economic stakes in whaling have atrophied with the rusting of the whaling fleets.[89] Malaysia's explicit rejection of a deforestation convention at the United Nations Conference on Environment and Development provides another illustration of this dynamic. In short, states tend to reject a new norm of sovereignty when its sole rationale lies in moral discourse.

PROPOSITION 3.   *Moral discourse leads those states that do not accept the underlying principled beliefs of that discourse to reject new norms of sovereignty unless acceptance of those norms is induced through more direct, material incentives.*

This case highlights how the interaction between scientific consensus and principled beliefs influences state willingness to accept new norms of sovereignty. To vastly oversimplify, environmental protection finds support in two different philosophies: a "deep" ecology committed to an epistemologically unconditioned belief in limiting human exploitation of nature's intrinsic values, and a "shallow" ecology committed to an epistemologically conditioned belief in limiting human activities when science demonstrates that those activities threaten nature's instrumental value to humans.[90] The principles and logic of a shallow ecology make support for a policy that we should not kill whales contingent on scientific consensus regarding how human behavior influences the various species of whales. Scientific evidence that an activity causes harm will produce support for regulation, while evidence that exonerates a previously suspect activity will undercut such support. In contrast, deep ecology's moral principles dictate a policy of protecting whales (and nature in general) that is not contingent on scientific evidence. Indeed, scientific evidence becomes logically irrelevant.

The precautionary principle, which underpins several recent international environmental treaties, highlights the contingent rather than determinist relationship of science and policy. By urging policy *action* in response to levels of scientific uncertainty traditionally used to rationalize policy *inaction*, this new legal principle demonstrates that deriving policy from scientific knowledge requires implicitly or explicitly value-laden judgments about social goals and norms, what constitutes evidence and consensus, and how we should respond to uncertainty and risk. In short, how and whether people derive policy guidance from a particular set of causal beliefs and scientific facts varies depending on often implicit moral principles regarding the role of science and the relationship of humans to their environment.

PROPOSITION 4. *Scientific consensus demonstrating an activity's instrumental harm to nature will mask the divergence in basic principles underlying environmental concern, strengthening support for environmental protection policies. Scientific consensus demonstrating the absence of such harms will highlight this divergence, weakening support for environmental protection policies.*

Conclusions

Security and trade negotiations seek to design cooperative policies to deal with conflicts primarily involving politico-economic interests.[91] Negotiations on social issues, such as human rights, slavery, and trade in women and children, seek to design cooperative policies to deal with conflicts primarily involving normative values in which one side views these behaviors as illegitimate under any circumstances.[92] International environmental negotiations seek to design cooperative policies to deal with conflicts among and within politico-economic interests, normative values, and scientific knowledge, presenting a more complex interplay of these three sources of pressure for international policy change. The nature of the global environmental problems governments and their citizens face often requires states to resolve a metanegotiation about norms of sovereignty with respect to the conditions under which states must make and abide by collective, rather than independent, decisions regarding their behavior. States have already created many of the necessary de jure redefinitions of sovereignty by negotiating and signing various environmental treaties and establishing corresponding regimes.

The whaling case suggests that states will be more likely to create the corresponding de facto redefinition of sovereignty, i.e., abide by collective regime decisions even when incentives for independent decision making exist, when arguments rely on scientific discourse rather than power and interests or moral discourse. The whaling case leads us to ask whether other environmental regimes support the claim that causal belief–based arguments are more effective than interest-based or principled belief–based arguments in leading states to accept and abide by collective decisions that constrain behavior previously considered a sovereign right.

If the emergence of world civic governance, pressure from environmental NGOs, and other processes create the deep-seated changes to values and behavior necessary to stem the tide of anthropogenic environmental damage, then de jure changes in international legal norms of sovereignty can come after the more important de facto ones that benefit the environment have occurred.[93] Unfortunately, those processes are likely to take considerable time to come to fruition. In the meantime, efforts, even if unsuccessful, should be made to develop international legal norms for collective decision making in arenas where independent decision making is the practice. The task ahead is to determine how best to induce states to abide by such collective decision making processes. The history of whaling suggests that, at least in the short term, regimes that foster scientific discourse can contribute more to the legitimacy and practical application of such norms than those that do not.

Notes

1. I want to thank Karen Litfin, Mark Zacher, Elizabeth DeSombre, and the participants in the conference on Rethinking Sovereignty and Environment for helpful criticisms of earlier drafts of this paper.
2. Kathryn Sikkink, "Human Rights, Principled-Issue Networks, and Sovereignty in Latin America," *International Organization* 47 (Summer 1993): 440.
3. The distinction used throughout this paper between interest-based, causal belief–based, and principled belief–based discourse builds on the work in Judith Goldstein and Robert O. Keohane, "Ideas and Foreign Policy: An Analytic Framework," in Judith Goldstein and Robert O. Keohane, eds., *Ideas and Foreign Policy: Beliefs, Institutions, and Political Change* (Ithaca, N.Y.: Cornell University Press, 1993), and Sikkink, "Human Rights," 411–41.

4. IWC, *Text of the International Convention for the Regulation of Whaling, 1946, and Its Protocol* (Cambridge, England: International Whaling Commission, 1956): Article 5.

5. Ken Conca, "Rethinking the Ecology-Sovereignty Debate," *Millennium* 23 (January 1994): 1–11; Ruth Lapidoth, "Sovereignty in Transition," *Journal of International Affairs* 45 (Winter 1992): 330–31; 334; Sikkink, "Human Rights," 441.

6. Karen T. Litfin, "Sovereignty and Environment," chapter 1, this collection; Janice E. Thomson, "State Sovereignty in International Relations: Bridging the Gap between Theory and Empirical Research," *International Studies Quarterly* 39 (June 1995), 219; Daniel Philpott, "Sovereignty: An Introduction and Brief History," *Journal of International Affairs* 48 (Winter 1995): 357; and J. Samuel Barkin and Bruce Cronin, "The State and the Nation: Changing Norms and the Rules of Sovereignty in International Relations," *International Organization* 48 (Winter 1994): 107.

7. Stephen D. Krasner, "Westphalia and All That," in Goldstein and Keohane, *Ideas and Foreign Policy*, 260; Barkin and Cronin, "The State and Nation," 109; Stacy D. VanDeveer, "States, Seas, and Regimes: Who's Sovereign Now." Paper presented at 36th annual convention of the International Studies Association (Chicago, 1995): 4; and Gene M. Lyons and Michael Mastanduno, "International Intervention, State Sovereignty, and the Future of International Society," *International Social Science Journal* 45 (November 1993): 527.

8. I am indebted to Thom Kuehls for this insight.

9. VanDeveer, "States, Seas, and Regimes," 2.

10. The following policy history draws extensively on J. N. Tönnessen and A. O. Johnsen, trans. R. I. Christophersen, *The History of Modern Whaling* (Berkeley: University of California Press, 1982); Patricia Birnie, *International Regulation of Whaling: From Conservation of Whaling to Conservation of Whales and Regulation of Whale-Watching* (New York: Oceana Publications, 1985); M. J. Peterson, "Whalers, Cetologists, Environmentalists, and the International Management of Whaling," *International Organization* 46 (Winter 1992): 147–86; and David D. Caron, "The International Whaling Commission and the North Atlantic Marine Mammal Commission: The Institutional Risks of Coercion in Consensual Structures," *American Journal of International Law* 89 (January 1995): 154–74.

11. Birnie, *International Regulation of Whaling*, vol. 2, 77ff.

12. Ibid., 168.

13. Steinar Andresen, "The Effectiveness of the International Whaling Commission," *Arctic* 46 (June 1993): 109.

14. Ibid., 112.

15. Garrett Hardin, "The Tragedy of the Commons," *Science* 162 (13 December 1968): 1243–48. For two excellent explications of the problems facing common-pool resources, see Elinor Ostrom, *Governing the Commons: The*

Evolution of Institutions for Collective Action (Cambridge, England: Cambridge University Press, 1990); and Richard Cornes and Todd Sandler, *The Theory of Externalities, Public Goods, and Club Goods* (Cambridge, England: Cambridge University Press, 1986).

16. Peterson, "Whalers, Cetologists," 159 and 161; and Andresen, "Effectiveness of IWC," 110.

17. Tönnessen and Johnsen, *History of Modern Whaling*, 157.

18. Ibid., 491–92.

19. Ibid., 506 and 514.

20. Ray Gambell, "International Management of Whales and Whaling: An Historical Review of the Regulation of Commercial and Aboriginal Subsistence Whaling," *Arctic* 46 (June 1993): 99.

21. See the chart of quotas and catches in Peterson,"Whalers, Cetologists," 165.

22. Andresen, "Effectiveness of IWC," 112.

23. Ibid.

24. Birnie, *International Regulation of Whaling*, 210.

25. Ibid., 214.

26. Andresen, "Effectiveness of IWC,"

27. J. L. McHugh, "The Role and History of the International Whaling Commission," in William E. Schevill, ed., *The Whale Problem: A Status Report* (Cambridge, Mass.: Harvard University Press, 1974): 306.

28. Tönnessen and Johnsen, *History of Modern Whaling*, 619.

29. Peterson, "Whalers, Cetologists," 161.

30. Birnie, *International Regulation*, 227; and Sidney Holt and Nina M. Young, *Guide to Review of the Management of Whaling* (Washington, D.C.: Center for Marine Conservation, 1990); 4.

31. Tönnessen and Johnsen, *History of Modern Whaling*, 592ff.

32. Peterson, "Whalers, Cetologists," 161–64; McHugh, "Role and History of IWC," 311; and Steinar Andresen, "Science and Politics in the International Management of Whales," *Marine Policy* 13 (April 1989): 105.

33. Birnie, *International Regulation*, 321, 326, and 339.

34. Ibid., 350; Alexey V. Yablokov, "Validity of Whaling Data," *Nature* 367 (13 January 1994): 108; and Caron, "IWC and NAMMCO," 171.

35. Andresen, "Effectiveness of IWC," 110; and J. Scarff, "The International Management of Whales, Dolphins, and Porpoises: An Interdisciplinary Assessment," *Ecology Law Quarterly* 6 (1977): 366.

36. Birnie, *International Regulation*, 453 and 464; and Ray Gambell, "Whale Conservation—Role of the International Whaling Commission," *Marine Policy* 1 (October 1977): 305.

37. R. Michael M'Gonigle, "The Economizing of Ecology: Why Big, Rare Whales Still Die," *Ecology Law Quarterly* 9 (Fall 1981): 158. Like the

Mediterranean politicians described by Haas, conditions of uncertainty made economic interests more willing to accept scientific recommendations, in Peter M. Haas, *Saving the Mediterranean: The Politics of International Environmental Cooperation* (New York: Columbia University Press, 1990).

38. Andresen, "Effectiveness of IWC," 110.

39. Birnie, *International Regulation*, 422 and 434.

40. Ibid., 438n 122.

41. Tönnessen and Johnsen, *History of Modern Whaling*, 683–84; and Birnie, *International Regulation*, 556.

42. Gene S. Martin, Jr., and James W. Brennan, "Enforcing the International Convention for the Regulation of Whaling: The Pelly and Packwood-Magnuson Amendments," *Denver Journal of International Law and Policy* 17 (Winter 1989): 298.

43. Susan Geha, "International Regulation of Whaling: The United States Compromise," *Natural Resources Journal* 27 (Fall 1987): 931–40; Martin and Brennan, "Enforcing the International Convention," 293–315; and Dean M. Wilkinson, "The Use of Domestic Measures to Enforce International Whaling Agreements: A Critical Perspective," *Denver Journal of International Law and Policy* 17 (Winter 1989): 271–92.

44. McHugh, "Role and History of IWC," 306.

45. Birnie, *International Regulation*, 429, 471, and 503.

46. Martin and Brennan, "Enforcing the International Convention"; and Wilkinson, "Use of Domestic Measures."

47. Hardin, "Tragedy of the Commons."

48. Tönnessen and Johnsen, *History of Modern Whaling*, 750; and McHugh, "Role and History of IWC," 306 and 310.

49. Martin and Brennan, "Enforcing the International Convention," 298–99.

50. Peterson, "Whalers, Cetologists," 155 and 169.

51. Lynton Keith Caldwell, *International Environmental Policy* 2d ed. (Durham, N.C.: Duke University Press, 1990): 286; and cited in Peter J. Stoett, "International Politics and the Protection of Great Whales," *Environmental Politics* 2 (Summer 1993): 296.

52. Birnie, *International Regulation*, 613–14.

53. Peterson, "Whales, Cetologists," 177.

54. Birnie, *International Regulation*, 616.

55. Ibid.

56. Ibid., 485 and 507.

57. Peterson, "Whales, Cetologists," 170.

58. Ibid., 169; and Andresen, "Effectiveness of IWC," 114.

59. Birnie, *International Regulation*, 638; Mari Skåre, "Whaling: A Sustainable Use of Natural Resources or a Violation of Animal Rights?" *Environment*

36 (September 1994): 12–20 and 30–31; and Anthony D'Amato and Sudhir K. Chopra, "Whales: Their Emerging Right to Life," *American Journal of International Law* 85 (January 1991).

60. Patricia Forkan, cited in Peterson, "Whales, Cetologists," 170; and Andresen, "Effectiveness of IWT," 113.

61. Arne Kalland, "Management by Totemization: Whale Symbolism and the Antiwhaling Campaign," *Arctic* 46 (June 1993): 124–33; and Mats Ris, "Conflicting Cultural Values: Whale Tourism in Northern Norway," *Arctic* 46 (June 1993): pp. 156–63.

62. Marine Mammal Commission, *Annual Report—Calendar Year 1990* (Washington, D.C.: 1990): 131–32; Greenpeace, "Outlaw Whalers," *Greenpeace* (1990): 1; and Greenpeace, "Scientific Whaling," *Greenpeace* (1985).

63. Litfin, "Sovereignty and Environment," 4. Greenpeace, "Outlaw Whalers," 1.

64. For discussions of recent dynamics within the IWC, especially including Norway and Japan's responses to them, see Robert L. Friedheim, "Moderation in the Pursuit of Justice: Explaining Japan's Failure in the International Whaling Negotiations." Paper presented at the Kyoto Conference on Japanese Studies, 1994; Dylan A. MacLeod, "International Consequences of Norway's Decision to Allow the Resumption of Limited Commercial Whaling," *International Legal Perspectives* 6 (Spring 1994): 131–58; Peter G. G. Davies, "Legality of Norwegian Commercial Whaling under the Whaling Convention and Its Compatibility with European Community Law," *International and Comparative Law Quarterly* 43 (April 1994): 270–95; Martha Howton, "International Regulation of Commercial Whaling: The Consequences of Norway's Decision to Hunt the Minke Whale," *Hastings International and Comparative Law Review* 18 (Fall 1994): 175–93; Cliff M. Stein, "Whales Swim for Their Lives as Captain Ahab Returns in a Norwegian Uniform: An Analysis of Norway's Decision to Resume Commercial Whaling," *Temple International and Comparative Law Journal* 8 (Spring 1994): 155–85; Caron, "IWC and NAMMCO"; and Skåre, "Whaling."

65. Stein, "Whales Swim for Life," 169; and Peterson, "Whales, Cetologists," 182.

66. Caron, "IWC and NAMMCO," 160.

67. Ibid.

68. Gambell, cited in Caron, "IWC and NAMMCO," 162.

69. Ibid., 161.

70. Greenpeace, "A New Era for the IWC," *Greenpeace* (October-November-December 1991): 5.

71. Kazuo Shima, Japanese commissioner on the IWC, cited in Caron, "IWC and NAMMCO," 162.

72. Stoett, "International Politics," 298.

73. IWC Resolution, cited in Caron, "IWC and NAMMCO," 170.

74. Friedheim, "Moderation in Pursuit of Justice," 12.

75. Mark Meredith, "Iceland Quits World Whaling Body," *Reuters News Service*, 29 June 1992, Glasgow, Scotland.

76. "Six Ships Avoiding Protesters," *New York Times*, 5 July 1992; and Mark Meredith, "Norway Announces Resumption of Commercial Whaling," *Reuters News Service*, 29 June 1992, Glasgow, Scotland.

77. Alf Håkon Hoel, "Regionalization of International Whale Management: The Case of the North Atlantic Marine Mammal Commission," *Arctic* 46 (June 1993): 116–23

78. Caron, "IWC and NAMMCO," 165.

79. Ibid., 168 and 173.

80. Ibid., 165.

81. Sikkink, "Human Rights"; Kathryn Sikkink, "The Power of Principled Ideas: Human Rights Policies in the United States and Western Europe," in Goldstein and Keohane, *Ideas and Foreign Policy*, 139–70; and Goldstein and Keohane, *Ideas and Foreign Policy*.

82. Friedheim, "Moderation in Pursuit of Justice," 10.

83. Caron, "IWC and NAMMCO," 155.

84. Karen Litfin, personal communication, 1997.

85. For a discussion of the distinction between the appropriation and provision problems involved in common-pool resources, see Ostrom, *Governing the Commons*, 46–50.

86. Haas, *Saving the Mediterranean*.

87. Sikkink, "Human Rights," 437.

88. On how reactive resistance helps explain the failure of economic sanctions to alter behavior, see James Barber, "Economic Sanctions as a Policy Instrument," *International Affairs* 55 (July 1979): 376–78.

89. Friedheim, "Moderation in Pursuit of Justice," 22ff; and Herman Prager, "Herman Daly and Problems of Sustainable Development: The Case of Japanese Whaling," unpublished paper (Monroe, La., 1995): 8–9.

90. Arne Naess, "The Shallow and the Deep, Long-Range Ecology Movement: A Summary," *Inquiry* 16, no. 1 (1972): 95–100; and Edith Brown Weiss, *In Fairness to Future Generations: International Law, Common Patrimony, and Intergenerational Equity* (Dobbs Ferry, N.Y.: Transnational Publishers, 1989).

91. Causal beliefs and normative values have played some role in such arenas as nuclear and chemical weaponry. See Emanuel Adler, "The Emergence of Cooperation: National Epistemic Communities and the International Evolution of the Idea of Nuclear Arms Control," *International Organization* 46

(Winter 1992): 101–46; and Richard Price, "A Genealogy of the Chemical Weapons Taboo," *International Organization* 49 (Winter 1995): 73–103.

92. Sikkink, "Human Rights"; Sikkink, "Power of Principled Ideas"; and Ethan A. Nadelmann, "Global Prohibition Regimes: The Evolution of Norms in International Society," *International Organization* 44 (Autumn 1990): 479–526.

93. Paul Wapner, "Environmental Activism and World Civic Politics," *World Politics* 47 (April 1995): 311–40; and Thomas Princen, Matthias Finger, and Jack Mann, "Nongovernmental Organizations in World Environmental Politics," *International Environmental Affairs* 7 (Winter 1995): 42–58.

7

Sovereignty Reconfigured: Environmental Regimes and Third World States

Marian A. L. Miller

Introduction

Developing countries' sovereignty is already compromised by the nature and structure of the world economy, and it is likely to be further modified by the evolving environmental regimes. Environmental regime provisions modify the degree of authority, autonomy, and control that can be exercised by states. Consequently, they prompt changes in the legal and effective sovereignty of both industrialized and Third World countries. But, because of the place of the Third World countries in the world economy, the impact on them might be amplified. This chapter examines the impact of environmental regimes on Third World sovereignty by focusing on the regimes addressing stratospheric ozone depletion, biodiversity, and hazardous waste trade.

International regimes, by their very nature, modify the norms and practices of state sovereignty. In recognition of their common interests, states collaborate to form social institutions[1] that govern their actions. This agreement to collaborate suggests a ceding of some authority to the regime and shifts in autonomy and control in certain policy areas. This reconfiguration of authority, autonomy, and control is experienced by all states. However, the nature of environmental issues, as well as the structural position of the states, can mean a difference in how effective sovereignty is reconfigured for industrialized states and for Third World states. Although there is intragroup diversity, each group of states retains significant common experiences and interests, and often employs common negotiating strategies. Because of these common experiences, interests, and strategies, industrialized states and Third World states can be regarded as distinct analytical categories in this examination of sovereignty and environmental regimes.

Sovereignty refers to a state's legal right to determine the norms of behavior and conditions of life within it.[2] But internationally recognized legal status as a sovereign state does not necessarily reflect the government's ability to exercise authority, autonomy, and control. A state's sovereignty does not guarantee that it can provide its population with the requisite civil order or that it can control entrances into or exits from the country. Even when legal or formal sovereignty remains intact, *effective* sovereignty varies greatly among states and is being modified by global forces even for the most powerful states. Across the board, states' authority, autonomy, and control are being reconfigured by the increasing mobility of goods, information, and people; new transnational associations and global markets; and the proliferation of practices affecting the global environment. But changes in the practice of sovereignty have been more marked for the states of the Third World.

Globalization, Interdependence, and Sovereignty

Both economic and ecological globalization have been factors in the reconfiguration of sovereignty, decreasing the autonomy and control of state actors, especially in the Third World. The concept of a national economy is becoming meaningless as the various factors of production—money, technology, factories, and equipment—move easily across national borders. Buyers, sellers, and investors can often sidestep national policies, thus undermining a country's ability to ensure a particular social and economic order. Transnational corporations can significantly reduce states' control over activities within their own territory. Since they control the bulk of world trade and investments,[3] they retain particularly effective means of reducing Third World states' options in a wide variety of policy areas, including industry, labor, environment, and taxation.

Intergovernmental organizations (IGOs) also undercut states' authority and control, but Third World states are particularly vulnerable. Their interactions with IGOs such as the General Agreement on Tariffs and Trade (GATT), the International Monetary Fund (IMF), and the World Bank are illustrative; these organizations reflect the structural advantage of the industrialized states. When these international economic organizations were created after World War II, their major purpose was to facilitate interactions among the industrialized countries. The IGOs are still run by industrialized states; and their decision-making and voting procedures are still weighted toward these more powerful states. These organizations provide avenues for intrusion into the economic life of developing countries, as decisions at the international level diminish states' control over their economies and decrease their development options. This can be seen when GATT provisions challenge the right of states to use import or export controls to protect and conserve fisheries and forests; when the shock treatments of IMF structural adjustment programs cause severe socioeconomic dislocations,[4] and when the World Bank funds megaprojects that result in environmental stress, disruptions of local ecosystems, and the displacement of thousands of people.[5]

State sovereignty is being modified not only by the global economy, but also by global environmental problems. No country can escape the scope of problems such as the damaged ozone layer and the changing climate. Because these challenges are global in scope, solutions require the cooperation of all state actors. A few powerful actors cannot solve these problems by themselves; the cooperation of those states traditionally viewed as weak is required. The very nature of global environmental problems can create new power dynamics between developed and developing countries, thereby changing the way in which gradations of power among countries are perceived. When biosphere issues are considered in a commons context, any actor has the ability to subtract from the welfare of the others. This provides Third World countries with a potential basis for leverage.[6] *how?*

Transboundary environmental issues also modify the exercise of sovereignty. Damage to rivers, seas, forests, airsheds, and animal and plant life often transcends national boundaries; solutions thus require the cooperation of actors in multiple states. Clearly, a state can no longer ensure the well-being of its citizens by focusing only on issues within its own borders, and it may no longer be allowed to carry on activities within its own borders with impunity, if those activities are seen as damaging to the well-being of other states.

Although states have generally been unwilling to give up decision-making autonomy in order to obtain security from common threats, the decrease in autonomy and control resulting from ecological interdependence has caused them to reconsider. In the process of managing these problems, some authority has been shifted from state institutions to international and global institutions. While little can be done to prevent the erosion of sovereignty inherent in commons or transboundary environmental problems, the kind of institutions created to address the problems can result in differentials in the modification of sovereignty across nations. These institutions can either amplify or mitigate the relative leakage of sovereignty for Third World states.

While economic and ecological globalization both result in interdependence, they produce two different kinds of interdependence. Because it is a consequence of the existing economic and social system, economic interdependence is a constructed type of interdependence.[7] The globalization of resources, capital, and labor has produced a growing network of intergovernmental and transgovernmental linkages. Consequently, all global actors are, to a greater or lesser degree, more sensitive to changes in others and more vulnerable to external circumstances.[8] This results in reduced autonomy and control for state actors. Although the multiple linkages are most pronounced among the industrialized countries, major intergovernmental organizations (IGOs) and nongovernmental organizations (NGOs) have been instrumental in integrating Third World countries into the global economy. Because of the power asymmetry between the Third World states and the industrialized states, the sensitivity and vulnerability that are, the consequences of an interdependent world are more likely to be reflected in the Third World as costs rather than as benefits. While this constructed interdependence is likely to be a long-term feature, it is not a permanent, unchanging characteristic of the world system.

On the other hand, ecological interdependence is perceived as a naturally occurring interdependence, since all of humankind has to operate within the constraints of the biosphere. In a system in which there is economic interdependence, economic asymmetry can be exploited to benefit the more powerful actors. However, a general perception of ecological interdependence places constraints on both the strong and the weak.

In the arena of international environmental politics, the economic differences between the Third World and the industrialized states result in a divergence of short-term interests. But these interests converge in the long term, since these states all have a common interest in the habitability of the earth. Because of the constraints of the biosphere, a perception of interdependence means a diffusion of power among all actors in the world system. As a result, Third World states have the potential to deny the industrialized states their environmental objectives. On particular issues, small groups of developing countries can form effective blocking coalitions. Because together they have the potential to accelerate environmental degradation, their demands have to be considered.

Interdependence is enhanced when a common property resource is involved. In that case, potential beneficiaries are difficult to exclude, and each user can subtract from the welfare of the other users.[9] Two of the three regimes under study deal with common property resource issues. In the case of the ozone layer protection regime, the resource is held in an open-access arrangement, where everyone has the ability to affect the atmosphere. Since each user is capable of subtracting from the welfare of the others, there is likely to be a perception of interdependence and vulnerability. Biodiversity is also a common property resource, but it is held in a mix of property arrangements: open-access, private property, communal property, and state control. The third regime, dealing with the hazardous waste trade, addresses a transboundary problem rather than a commons problem. Property and access arrangements vary with each environmental problem, and they affect how states perceive their interdependence and vulnerability.[10] Consequently, property and access issues have implications for the leverage possessed by the Third World in international environmental negotiations, and for the degree to which they can slow the shift of effective sovereignty.

Stratospheric Ozone Regime

The earth's ozone shield is being depleted by chemicals, such as chlorofluorocarbons (CFCs) and halons, which are important components of the industrialization process. Most CFC production and consumption has taken place in the industrialized world; therefore those countries gain the greater benefit from these chemicals. But since their use damages the ozone shield that protects the global community, the industrialized countries are able to transfer some of the cost of their industrialization to the other members of the global society. Because of the nature of the problem, no country can act individually to prevent damage to itself or any other.

In 1978, the United Nations Environment Programme (UNEP) established a committee to examine ozone layer depletion, a process that culminated in a framework convention, the Vienna Convention for the Protection of the Ozone Layer, which was adopted in 1985. Two years later, this was supported by the Montreal Protocol on Substances That Deplete the Ozone Layer, requiring signatory governments to regulate production and consumption of CFCs and halons. The protocol entered into force in 1989, and was amended in 1990 to include a larger group of chemicals. The amendment entered into force in August 1992. In November of that year, the parties to the convention agreed to accelerate the phaseout schedule. Developed countries would phase out their use of CFCs by January 1, 1996. In 1995, the Montreal Protocol was further modified to tighten restrictions on hydrochlorofluorocarbons (HCFCs) and methyl bromide.

Because of the nature of the problem, ozone layer depletion contributes to a reconfiguration of effective sovereignty for all states. The problem underscores a lack of state control, since it demonstrates states' inability to establish a particular internal environmental order. The characteristics of the atmosphere clearly limit states' options. Not only is the atmosphere a common property resource, with each user capable of subtracting from the welfare of others, but it is also in an open-access arrangement; consequently, it is impossible to control others' access to the atmosphere. The interdependence of all users, therefore, necessitates a particular kind of *sovereignty bargain*, as Karen Litfin has laid out the term in the introduction to this volume. In this bargain, all actors must cede some degree of autonomy in exchange for increased collective control over ozone-depleting substances. Both the nature of the problem and the scope of the solution have implications for legal as well as effective sovereignty.

In the initial phase of the regime evolution process, Third World countries were primarily observers. They did not see themselves as major stakeholders since industrialized countries were the major producers and consumers of ozone-depleting chemicals. The issue was defined and the agenda set by individuals, organizations, and nations of the industrialized world. A transnational epistemic community, drawn primarily from industrialized countries, played a major role in the definition of the issue. The major CFC producers were First World corporations, which were aware that the issue could be defined in such a way as to be detrimental to their economic interests.

After 1987, the Third World states realized that they had significant stakes in the regime development process. A major agenda item was the phaseout of ozone-depleting chemicals, and Third World states were concerned about the implications for their own development. They realized that the targeted chemicals had made significant contributions to industrial growth in developed countries, and they were interested in retaining access to the technologies, or, failing that, in getting financial assistance to help them acquire ozone-friendly technologies.[11] For their part, developed states were aware that as developing states industrialized they would represent an increasingly significant threat to the ozone layer. The position of China and India is illustrative. Although these two countries have about 40 percent of the world's population, they are responsible for only about 5 percent of CFC consumption.[12] CFCs are in demand in developing countries for a variety of uses, including cooling. As the use of refrigerators and air conditioners spreads in India and China, the need for CFCs or chemicals with similar properties will grow significantly. So there is the potential for developing countries to contribute significantly to ozone layer damage.

Because the developing states' cooperation was so important to the success of any management regime, they were able to bargain for concessions. One major concession was a modification of the chemical phaseout schedule. Developing countries would be allowed to increase their CFC use for the first decade of the agreement. As the regime evolves, the pattern of providing a grace period for developing states continues. For example, as a result of the strengthening of the Montreal Protocol in 1995, developed countries will phase out HCFCs by the year 2020 and methyl bromide by 2010. But developing countries have a 2040 phaseout date for HCFCs, and they have agreed to freeze methyl bromide use at average 1995–98 levels by 2002. In 1997, the methyl bromide phaseout date was moved from 2010 to 2005 for industrialized countries, while developing countries agreed to a 20 percent reduction by 2005 and a phaseout by 2015.

Not all developing countries will use their entire grace periods, though. Consumption of ozone-depleting substances is falling in some developing countries,[13] and more than one-third of the countries in the developing group have committed to phasing out CFCs earlier than required.[14] Still, the use of ozone-depleting substances in developing countries is increasing, because of rapidly rising consumption in countries with large populations such as China and India.[15]

Another concession was the establishment of a fund to assist developing countries in their transition to more ozone-friendly technologies. Initially, some industrialized countries opposed the fund. There was particularly strong resistance from the United States, but it finally gave in under pressure from both industrialized and Third World states. Industrialized country actors knew that Third World countries, if they chose to industrialize with the old CFC technology, would deny the regime its objective. Scientific evidence suggested that the consequences of a failed regime would include damage to human health, agriculture, and ecosystems. This understanding contributed to the perception of interdependence held by the developed states.

The concession led to the establishment of the Multilateral Fund. The fund is managed by an executive committee made up of seven representatives each from the developing and developed country groups. The management structure recognizes that these groups may have different priorities. The committee tries to operate by consensus, but when this is not possible, a two-thirds majority from each group is required.[16] The fund has its share of problems. Its resources are inadequate, especially since money committed by developed countries is sometimes withheld. There are also delays in disbursement and implementation.[17]

The resulting ozone layer protection regime entails a sovereignty bargain whereby states agree to a loss of autonomy and authority in exchange for greater collective environmental control. For while the ozone regime decreases formal sovereignty for all parties, this transfer of formal sovereignty is likely to yield mixed results in terms of effective sovereignty. On the one hand, it places constraints on states' industrial options, but, on the other hand, it enhances their ability to improve environmental conditions. Because of the interdependence resulting from the atmosphere's common property status, Third World countries were able to bargain for concessions that would slow the drain of effective sovereignty that is a consequence of their position in the world system. Reflecting the interdependence of all global actors, the regime discourages the indefinite transfer of environmental costs that would result from industrialized countries' continued use of CFCs and similar chemicals. It therefore supports developing countries' efforts to establish a particular environmental order within their jurisdictions.

Biodiversity Regime

Destruction of biodiversity includes depletion of genetic diversity, species extinction, and disruptions to ecosystem areas such as oceans and forests. Although the loss of biodiversity has been defined as a commons issue, it is somewhat different from the ozone depletion problem. While the atmosphere is held in an open-access regime, biodiversity is held in mixed property arrangements. All four types of property arrangements are relevant here: open-access, private property, communal property, and state control. For example, fish or whales traversing the high seas are in open-access arrangements; the flora and fauna on the family farm are private property; shared community fishing grounds are communal property; and the national parkland is under state control. These mixed property arrangements for biodiversity diminish perceptions of interdependence and vulnerability. Another complicating factor is the concentration of biodiversity in the Third World. Tropical forests hold "at least 50 percent and perhaps 90 percent of the world's species."[18] While states have formal sovereignty over the portion of biodiversity within their jurisdictions, there is the countervailing perception that all biodiversity is a part of the common heritage of mankind. The Third World and industrialized states' conflicting perspectives on biodiversity were played out in the negotiations that led to the adoption of the Convention on Biological Diversity. The convention entered into force on December 29, 1993.

The developed countries initially focused on conservation; however, Third World states were able to broaden the scope of the agenda. The industrialized countries wanted to protect biological diversity and to be assured of continued access to Third World biological resources. There are considerable economic interests at stake in genetic resources, since gene technology has significant implications for sectors ranging from agriculture to medicine. For their part, the Third World states had their own demands: sovereignty, access to technology, and financial assistance. These countries took the position that biodiversity was a resource like oil or lumber, and that they had the right to control access and get reimbursement. They also wanted access to technology, including the biotechnology developed from the use of their resources and technology to help them conserve their biodiversity. And, finally, they wanted financial help to make the transition to sustainable development.

Their bargaining tactic was to link accession to the treaty and access to their biological resources to financial and technical assistance. Their efforts yielded mixed results. Their sovereignty was formally recognized in the treaty;[19] however, they were less successful with regard to technology transfer, and it is difficult to assess their success regarding financial assistance. The provisions for technology transfer and the sharing of research were substantially modified by language addressing intellectual property rights and patents. The convention provided for the transfer of technologies under fair and favorable terms, including concessional and preferential terms; however, if patents and intellectual property rights are involved, any transfer has to be consistent with the protection of those rights.[20] At the third meeting of parties to the convention, in November 1996, developing countries noted the lack of activity in this area, emphasizing that little had been done to transfer technology.[21]

On the financial side, the agreement included a commitment for developed country parties to fund the "agreed full incremental costs" incurred by developing countries in meeting the obligations of the convention.[22] Consequently, the funding arrangement is an important concern for developing countries. The designated funding mechanism for the convention is the Global Environment Facility (GEF). It is jointly managed by the World Bank, United Nations Development Programme, and the United Nations Environment Programme. Decisions are made on a double-majority voting system that allows donors effective veto power.

The GEF is an interim financial mechanism. From the beginning, Third World countries questioned whether it would effectively address their interests. So far, they have resisted pressure from donor countries to accept the GEF as the convention's final financial mechanism. They prefer a mechanism that is not dominated by the larger donor governments, and one that is under the control of the convention's Conference of Parties. Such a mechanism would oversee not only the disbursement of money from donor governments, but also the allocation of funds from users of biological resources, who are required to share the benefits of such use.[23] At the third conference of parties, it was agreed that the GEF would continue to serve the biodiversity convention on an interim basis, and a review of the financial mechanism would take place at the fourth Conference of Parties in May 1998.[24]

Although Third World countries did not get all that they wanted, the language regarding sovereignty, control of access, and the sharing of the profits of technology might indicate a shift in norms. This change is reflected in the contract between a Costa Rican nonprofit research institute and Merck and Company. Merck has agreed to pay the institute $1.1 million, as well as royalties from any product developed, in return for plant, insect, and microbe samples from Costa Rica.[25] Because so much biodiversity is within the sovereign control of Third World countries, individual countries can take formal action to effectively protect biodiversity within their borders. But the mere passing of statutes will control neither the predation of local capital nor the intrusion of global economic forces. To a large extent, global biodiversity is being privatized as patent lawyers and transnational corporations assert property rights to genetic material. These moves have been substantially supported by GATT's provisions addressing intellectual property rights. One outcome of the Uruguay Round of GATT was a call for the international harmonization of property rights legislation. Signatory nations are required to introduce a U.S.-style regime of intellectual property rights. This requirement will push local knowledge from the social sphere of common rights to the commercial sphere of free global access.[26]

Consequently, although the common property status of biodiversity allowed Third World countries to bargain successfully for formal recognition of their right to control access to their biological resources, the prospects for effective sovereignty have been mixed. In some cases, countries have been able to establish, and profit from, effective economic ownership, but the intrusion of the global economy is rapidly undermining efforts at local control. Absent is the strong perception of interdependence and vulnerability which would mitigate the continued shift of sovereignty from the Third World states. For industrialized countries, the perception that others can subtract from their welfare is not as significant with biodiversity as it is in the case of ozone layer depletion. In addition, Third World countries' formal control of access to significant biological resources is not a major barrier to global economic interests. In this case, therefore, economic globalization is able to counter the effects of ecological globalization and prevent the gains in formal sovereignty from being translated into enhanced effective sovereignty for Third World states.

Hazardous Waste Trade Regime

Approximately 90 percent of the world's hazardous waste is generated by industrialized countries.[27] In many of these states, landfills and incinerators have been the principal, waste-disposal options, but stricter regulations and decreasing disposal options have prompted hazardous waste producers to seek disposal sites in other countries. Although only about one-fifth of the global trade in wastes goes to developing countries,[28] it is a major concern because these countries often lack the technology or administrative capacity to handle the wastes safely. In addition, the people in the recipient countries are often poorly informed about the toxicity of the waste materials. This problem received considerable media attention in the late 1980s, when there was a spate of news stories about waste dumping in the Third World. The Cairo Guidelines and Principles for Environmentally Sound Management of Hazardous Wastes, adopted in 1987, represented the first attempt to address the matter from a global perspective. This was followed by the Basel Convention on the Control of Transboundary Movements of Hazardous Wastes and Their Disposal, which was adopted in March 1989 and entered into force in 1992. The convention was strengthened in March 1994 when contracting parties agreed to a complete ban on the shipping of hazardous wastes from OECD (Organization for Economic Cooperation and Development) to non-OECD states, to be fully in effect by December 31, 1997.[29] In September 1995, delegates to the third Conference of Parties of the Basel Convention agreed by consensus to amend the convention to reflect the ban.[30] The amendment will enter into force after its ratification by three-fourths of the country parties. As of February 1998, the amendment had sixteen ratifications; forty-eight more states must ratify it for it to become international law.

Third World states and environmental NGOs played a major role in publicizing the problem, but the Third World countries were initially unsuccessful in defining the issue to reflect their own interests. Some Third World states wanted to define the problem as the trade itself. If the problem is defined in that way, the logical solution would be the complete banning of the hazardous waste trade. The industrialized states, however, insisted that the problem should be defined as the regulation of the trade. As the bargaining proceeded, developing countries continued to press for a complete ban, while the developed countries were arguing for loose controls. Although developing countries and environmental NGOs were part of a high-profile media campaign during this stage, at the conclusion of the negotiations on the Basel Convention in 1989 they still had not succeeded in redefining the problem.

The issue of sovereignty was paramount here, since the management or banning of the trade involved the ability to control what entered a country. The first major effort at addressing the problem (the Cairo Guidelines) paid attention to the issue of formal sovereignty, since its provisions included the following: prior notification of the receiving state of any export; consent by the receiving state prior to export; and verification by the source state that the receiving state's disposal requirements are at least as stringent as those of the source state. But these declarations were soft-law instruments and therefore nonbinding. In any case, the declarations did not affect the movement of hazardous waste.

Developing countries were critical of the fact that the Cairo Guidelines did not ban the movement of hazardous wastes. Later on, they criticized the Basel Convention for the same reason. The convention set up an informed-consent regime, which would require waste exporters to notify their governments of any exports and to notify receiving countries of any shipments prior to their arrival.[31] The Third World countries continued to press for a global ban because they did not believe unilateral action would be sufficient to stop the trade. They realized that their effective sovereignty would be compromised, since they could not feasibly screen all questionable imports. However, in light of the weak Basel Convention, developing countries took the only real options available to them: They instituted unilateral and regional bans, even though it was not clear how well these bans could be implemented. At the very minimum, the bans did serve as a statement of preferred norms of behavior. In 1994, as a result of continued pressure on industrialized countries as well as the unilateral bans imposed by more than ninety developing countries, the parties to the Basel Convention agreed to a total ban on the movement of hazardous wastes from OECD to non-OECD countries. At a 1995 meeting of the Conference of Parties, the convention was amended to reflect this ban. For wastes intended for final disposal, the ban was effective immediately; for wastes intended for recycling, the ban is to take effect at the end of 1997.

This agreement will be difficult to enforce. A major concern is whether the states can successfully monitor the activities of powerful global corporations. In a globalized economy, corporations seek the lowest-cost waste sinks, many of which happen to be in the Third World. An even greater concern is whether the major industrialized states really want to monitor the waste trade. The United States, which generates approximately 80 percent of the world's total output of hazardous wastes,[32] is not a party to the convention. In addition, Australia, Austria, Canada, and Germany, parties to the convention, claimed that the 1994 ban was not legally binding.[33] The European Commission supported their position. In the negotiations leading up to the 1995 amendment of the Basel Convention, Australia, the United States, Canada, Japan, and New Zealand worked to weaken the ban by changing the language and questioning definitions of hazardous waste. In addition, after the ban's adoption, Australia and Canada talked about the possibility of circumventing the ban by way of bilateral agreements.[34] Developed country interests have also moved to weaken the regime through changes in OECD and European Union regulations.[35] In this way, they have modified their definition of the hazardous wastes that should be handled under the Basel Convention regulations.

In the period before the 1994 agreement on the ban, the Third World coalition held together under intense pressure. The pressure continued in the period leading up to the 1995 amendment, as some OECD trade representatives told Third World states that the ban would have a negative impact on their industrial development. The continued pressure was successful in creating a rupture in the non-OECD group as delegates from Brazil, India, and South Korea joined OECD delegates' efforts to undermine the ban.[36] However, the opposition group was not able to undermine the significant majority support for the ban among both OECD and non-OECD states.

Unlike the ozone depletion and biodiversity issues, hazardous waste trade is considered a transboundary issue and is not perceived to have the same commons implications. The trade also had a higher relative salience for the importing states than for the exporting states. The fact that the hazardous waste trade transferred environmental and economic costs to the importers, while the exporters—primarily the developed states—received environmental, political, and economic benefits, meant that it was difficult to galvanize industrialized states' support for significant change. While the importing states also stood to earn economic advantages from taking the waste, in many cases they preferred not to bear the associated environmental and political costs.[37]

The Basel Convention is encouraging new norms of behavior, and the Basel ban represents a hard-fought victory for Third World states. But the realities of global power will determine the regime's future. Those who would benefit from violating the ban are also those who are in positions of economic power. In addition, the United States is still not a party to the convention and therefore not required to observe its rules. These circumstances will determine the evolution of the regime. The parties to the convention have yet to establish a mechanism or procedure for monitoring and compliance. This was to be discussed in October 1997, at the fourth conference of parties.[38] Monitoring and compliance mechanisms notwithstanding, the OECD and global corporations are in a position to set the terms for regime transformation. Because hazardous waste is a transboundary rather than commons issue, self-interest is likely to predominate over broader community interests.

In the hazardous waste case, formal sovereignty was never under threat. As a matter of fact, both the Cairo Guidelines and the Basel Convention underscored the right of states to control access to their territories. But for the Third World these acknowledgements of formal sovereignty were negated by the nature of the global economy in which powerful actors were motivated and enabled to shift costs. The capacity of Third World states to stop or slow this shift was very limited; their control and autonomy were effectively challenged.

Because of the transboundary nature of the problem, there were no strong perceptions of interdependence to constrain developed country actors. So while formal sovereignty was unchallenged, the effective sovereignty of Third World states was, and continues to be, undermined.

Conclusion

An examination of these three cases suggests that Third World countries are better able to slow the leak of their effective sovereignty when the matter under debate is a common property resource concern than when it is otherwise. Like all state parties, Third World states cede some authority to regimes. But under commons circumstances, they are able to retain a significant degree of autonomy and control. Other actors are more willing to make compromises to get Third World cooperation, as was the case with the ozone layer protection regime.

The ozone regime underscores the importance of interdependence, which is a major consequence of common property circumstances. Because the perception of interdependence is greatest with respect to ozone depletion, Third World countries were able to exercise the greatest influence on this regime. Developed states were well aware of the damage that could result from Third World noncooperation. The Vienna Convention and the Montreal Protocol reconfigured sovereignty as all state parties ceded some authority and autonomy to the regime; however, developing countries were able to negotiate some mitigation of the erosion of their effective sovereignty. This would allow them greater environmental control and it would permit them to retain certain industrialization options.

Biodiversity issues presented a different context. Although biodiversity is a common property resource, not all of it is held in an open-access arrangement. Hence there was not the same perception of interdependence and vulnerability engendered by ozone depletion. While industrialized states might experience some deprivation, they do not see themselves suffering any significant damage in the short or medium term. The resultant regime supports legal sovereignty, but the growing attention to biotechnology and the adherence to intellectual property rights highlight the economic vulnerability of Third World countries.

Most of them lack the technological ability to exploit or safeguard their biological resources. In addition, the bulk of related patents are held by industrialized world entities. As a result of these factors, industrialized world economic interests drive biodiversity policy. Although conservation is not forgotten, it is a secondary factor, outweighed by global economic forces. Consequently, Third World countries' autonomy and control are weakened relative to the industrialized world. This is largely a function of the global economy, which aggravates, deepens, and exacerbates their area of vulnerability.

In the case of the hazardous waste trade regime, the Third World had no bargaining chip. Developed states would be neither damaged nor deprived by retention of the status quo; nor was there any perception of interdependence on their part. Because hazardous waste trade is a transboundary, rather than a common property resource, concern, developing countries could not offer developed countries environmental goods in order to get their cooperation to ban the trade. The ban would, in fact, deprive the exporting states of environmental goods. And very few Third World countries are in the position to offer developed countries scarce economic goods in exchange for cooperation. While the resulting regime supports legal sovereignty, this can be effectively ignored by major state and corporate actors.

Both the biodiversity and the hazardous waste trade regimes allow the Third World countries formal control of access, thereby reinforcing legal sovereignty. But because of the nature of the global economy, along with developing countries' administrative problems, the regimes do not slow the drain of effective sovereignty.

Another difference between the ozone layer regime and the regimes for biodiversity and hazardous waste trade is the complexity of regime implementation. Since the ozone regime's major requirement is for technological change in the industrial sector, agreements by nation-states, backed up by action in the relevant economic sector, will go a long way toward addressing the environmental concerns. However, environmental problems such as the depletion of biodiversity and the hazardous waste trade depend on substantial socioeconomic change and on cooperation at the global, regional, national, and community levels. In a global context where individuals and firms make their own arrangements independent of the interests or intent of states, there are many permutations that would allow potential transfers of sovereignty from Third World states.

This study suggests that, in spite of the differences in economic and political power, common property resource issues might allow Third World countries to exercise some modest influence in environmental regime formation and to slow the shift of authority, autonomy, and control to regimes or more powerful actors. Consequently, Third World states might be able to exercise similar influence with regard to the climate change regime, since it addresses a common property resource held in an open-access arrangement. Property and access relations, because of their impact on perceptions of vulnerability and economic interests, are likely to determine the further transformation of these regimes. However, in the absence of the perceived interdependence that is the consequence of common property circumstances, the industrialized world is likely to continue the transfer of economic and ecological costs to the developing world, thus accelerating the drain of effective sovereignty from Third World states.

Notes

1. See Oran R. Young, *International Cooperation: Building Regimes for Natural Resources and the Environment* (Ithaca, N.Y.: Cornell University Press, 1989): 12.

2. Seyom Brown, *New Forces, Old Forces, and the Future of World Politics: Post–Cold War Edition* (New York: HarperCollins, 1995): 2.

3. "The Power of Transnationals," *The Ecologist* 22, no. 4 (July-August 1992): 159.

4. For analysis of two countries' experience with structural adjustment programs, see Karen Hansen-Kuhn, "Sapping the Economy: Structural Adjustment in Costa Rica," *The Ecologist* 23, no. 5 (September-October 1993): 179–84; and Michel Chossudovsky, "India under IMF Rule," *The Ecologist* 22, no. 6 (November-December 1992): 271–75.

5. For analysis of several World Bank projects, see Bruce Rich, "The Multilateral Development Banks, Environmental Policy, and the United States," *Ecology Law Quarterly* 12 (1985): 685–88; Stephan Schwartzman, *Bankrolling Disasters: International Development Banks and the Global Environment* (San Francisco: Sierra Club, 1986); and Bank Information Center, *Funding Ecological and Social Destruction: The World Bank and International Monetary Fund* (Washington, D.C.: Bank Information Center, 1990).

6. Marian A. L. Miller, *The Third World in Global Environmental Politics* (Boulder, Colo.: Lynne Rienner Publishers, 1995): 10.

7. Ibid.

8. See Robert O. Keohane and Joseph S. Nye, *Power and Interdependence: World Politics in Transition* (Boston: Little Brown, 1977).

9. F. Berkes, D. Feeny, B. J. McCay, and J. M. Acheson, "The Benefit of the Commons," *Nature* 340 (July 13, 1989): 91–93.

10. Miller, *Third World in Global Environmental Politics*, 61.

11. Ibid., 73.

12. World Resources Institute, *World Resources 1992–93* (New York: Oxford University Press, 1992): 152.

13. Hilary F. French, "Ozone Response Accelerates," in Lester R. Brown et al. eds., *Vital Signs 1997* (New York: Norton, 1997): 102.

14. Multilateral Fund Secretariat, "Country Programme Summary Sheets," (Montreal: October 1996), cited in French, "Ozone Response Accelerates," 102.

15. French, "Ozone Response Accelerates," 102–3.

16. "Working Together to Protect the Ozone Layer: The Multilateral Fund for the Implementation of the Montreal Protocol," *UNEP IE OzonAction Programme: Multilateral Fund* (October 1997). Available: http://www.unepie.org/ozat/mf.html. 12.

17. "Ozone Protection Issues That Need Urgent Attention," Greenpeace Position Paper Prepared for the 9th Meeting of the Parties to the Montreal Protocol, September 1997, Montreal Canada. Available: http://www.greenpeace.org/~ozone/greenfreeze/env_imp/5protect.html. 7 October 1997.

18. World Resources Institute, *World Resources 1992–93*, 130.

19. United Nations Environment Programme, *Convention on Biological Diversity*, Article 15.

20. Ibid., Article 16.

21. "Third Session of the Conference of the Parties to the Convention on Biodiversity: 4–15 November 1996," *Earth Negotiations Bulletin* 9, no. 65, published by the International Institute for Sustainable Development. Available: http://www.mbnet.mb.ca/linkages/download/asc/enb0965e.txt. 7 October 1997.

22. UNEP, *Convention on Biological Diversity*, Article 20.

23. Alistair Graham, "Biodiversity Convention Begins to Stir," Community Biodiversity Network. Available: http://www.peg.apc.org/~bdnet/bulletin/cop2rpt.htm. 7 October 1997.

24. "Third Session of the Conference of the Parties to the Convention on Biological Diversity."

25. World Resources Institute, *World Resources 1992–93*, 138.

26. Vandana Shiva and Radha Holla-Bhar, "Intellectual Piracy and the Neem Tree," *The Ecologist* 23, no. 6 (November–December 1993): 227.

27. Gareth Porter and Janet Welsh Brown, *Global Environmental Politics* (Boulder, Colo.: Westview Press, 1991), 85. Also Greenpeace, *Toxic Trade Update* 5, no. 2: 5–6 (1992), cites Mostafa Tolba as saying that 98 to 99 percent of all hazardous wastes is produced by the developed countries, if this group is seen as including the Commonwealth of Independent States.

28. Porter and Brown, *Global Environmental Politics*, 85. Other estimates have placed this at 10 percent. This is the estimate of the UNEP Environmental Law and Institutions Unit, cited in Greenpeace International, "Annotations by Greenpeace International on the Agenda of the Meeting, Prepared for the First Conference of Parties to the Basel Convention 30 November–4 December 1992, Piriapolis, Uruguay," 2.

29. Paul Lewis, "Western Lands, except U.S., Ban Export of Hazardous Waste," *New York Times*, March 26, 1994.

30. Marcelo Furtado, "Basel Ban Here to Stay," *International Toxics Investigator* 8: 1 (First Quarter, 1996): 7.

31. Porter and Brown, *Global Environmental Politics*, 86–87.

32. Mostafa Kamal Tolba and Osama A. El-Kholy, *The World Environment 1972–1992* (London: Chapman and Hall, 1992): 264; also, Charles E. Davis (*The Politics of Hazardous Waste* [Englewood) Cliffs, N.J.: Prentice-Hall, 1993]: 4) cites a study by the Congressional Budget Office that concluded that approximately 266 million metric tons of hazardous wastes are generated in the United States annually.

33. Jim Puckett, "The Basel Ban Decision Explained," *Toxic Trade Update* 7, no. 1 (1994): 7–8.

34. Furtado, "Basel Ban Here to Stay," 7.

35. See Organization for Economic Cooperation and Development, *Decision of the Council Concerning the Transfrontier Movements of Wastes Destined for Recovery Operations* (Paris, April 6, 1992); also Paul Johnston, Ruth Stringer, and Jim Puckett, *When Green is Not*, Technical Note 07/92 (Amsterdam: Greenpeace International, 1992).

36. Furtado, Basel Ban Here to Stay," 7.

37. Miller, *Third World in Global Environmental Politics*, 134.

38. "Progress on Legal Issues Related to the Basel Convention," press release, UNEP, Nairobi, June 1997. Available: http://www.unep.ch/basel/sbc/pr697.htm. 2 October 1997.

8

Satellites and Sovereign Knowledge: Remote Sensing of the Global Environment

Karen T. Litfin

Introduction[1]

Environmental problems are not just physical occurrences; they are informational phenomena which are socially constructed through multiple struggles among contested knowledge claims. Access to and control over information, therefore, are crucial and controversial elements in environmental decision making. The gathering of environmental data is almost never a simple empirical exercise in sense observation, but is mediated by technologies of varying scale, scope, and sophistication. Those technologies, like any other, are more than simply neutral tools: They may be seen as "artifact/ideas" which embody political cultures.[2] Technologies only appear to be neutral because people generally assign a functional meaning to tools, rather than grasping the less visible logic of power, authority, and control that permeates them from their inception to their everyday usage. Artifacts should be viewed contextually, with different uses and meanings for different social groups, and should not be treated as objective facts.

Increasingly, the information that guides international environmental policy-making, both directly through treaties and regulations and indirectly through its impact on development activities, is being obtained through space-based instruments orbiting Earth on satellites. Earth remote sensing (ERS) can generate data on an enormous range of issues, including forest cover, the health of crops, atmospheric concentrations of many pollutants, drought conditions, crisis monitoring, resettlement of refugees, storm warnings, and the locations of resources ranging from drinking water, to petroleum and mineral deposits, to endangered species.[3] During the 1990s, approximately fifty Earth observation satellites will be launched by the space-faring nations of the world.[4] Although earth remote sensing (ERS) is indisputably a useful tool, it is also an artifact/idea whose political and social dimensions have been largely unexamined. Ironically, the very ubiquity of ERS capabilities, which gives it the semblance of the ultimate all-purpose neutral tool, may also impede serious analysis of the political culture embodied in it.

The implications of ERS for sovereignty are not straightforward; it is not a technology that can universally be said to either reinforce or erode modern practices associated with sovereignty as an institution. On the one hand, as a manifestation of "Big Science" and a supplier of information as a public good, ERS simulateously relies upon and supports the role of the state. As both Anthony Giddens and Michel Foucault have argued, although in different ways, surveillance technologies have been the basis for the state's administrative power throughout the modern era.[5] Indeed, surveillance provides the informational dimension of sovereignty, a dimension that is typically neglected in discussions of the state power and authority. Not surprisingly, the first (and, until recently, sole) users of satellite technology were the militaries of the two superpowers.

As the range of applications and the number of users has increased exponentially, in the past two decades, the relationship between satellite technology and state sovereignty has grown more complex. Today, users include multinational corporations, scientists, policymakers, grass roots environmental groups, and indigenous peoples. The multibillion-dollar industries of satellite communications and geographical information services (GIS) have dwarfed the military uses of satellites.[6] The loosely coordinated international global change research program, which relies primarily on satellite observations for its data, is likely to become the largest research project in human history, even with the current budget-cutting mood of many governments.[7] Information gathered through ERS will be applied by diverse sets of actors to some highly politicized purposes, including assigning responsibility for environmental degradation.

Is there a logic to satellite technology in general, and ERS in specific, with respect to sovereignty as a political institution? This chapter argues that, while the relationship is characterized by multiple, cross-cutting logics, certain traits and applications of ERS technologies are contributing to the reconfiguration of sovereignty. This is not to draw any grandiose conclusions about the abililty of satellite technology, in and of itself, to abrade or fundamentally alter the nation-state system. Rather, I focus on three attributes of ERS — transparency, globality, and its "high-tech" nature — and seek to grapple both inductively and deductively with their implications for some of the "bundles" that constitute the political institution of sovereignty. In particular, I examine the relationship of ERS to each of the following: the principle of territorial exclusivity, sovereignty bargains and international cooperation, popular sovereignty, and, the epistemic dimensions of sovereignty.

I argue that the relationship between ERS and sovereignty is multidimensional, in part because ERS technology seems to have emerged at the intersection between modernity and postmodernity. Although the term is hotly contested, the political meaning of *postmodernism* is usually taken to embrace two related tendencies: the diffusion of power along multiple capillaries and the proliferation of images and information.[8] If sovereignty is understood in terms of control and authority, the diffusion of satellite-generated data appears to be fostering multiple channels of control and authority, and therefore multiple sovereignties or micropowers, even as the technology owes its existence to large-scale state endeavors. Without making any sweeping predictions about the demise of sovereignty, one may infer that Earth remote sensing is a participant in the unfoldment of both of these apparently contradictory tendencies, and that the practices of sovereignty are undergoing revision in the late modern period, catalyzed by artifact/ideas manifested in specific technological forms.

Territoriality

Space Espionage and Extraterritoriality

As John Ruggie has argued, the principle of territorial exclusivity, an epochal development marking the end of the medieval era, has been the defining feature of the modern system of states. With the recent globalization of human activities, he claims, we are now witnessing the "unbundling of territoriality" and the "rearticulation of political space."[9] There is perhaps no form of technology better suited to exemplify these trends than ERS, which inherently erases territorial boundaries by virtue of the global scope of both its observations and its diffusion of information. While the transparency afforded by ERS no doubt undercuts the principle of territorial exclusivity, which is probably its greatest effect, ERS technology can also reinforce the modern practices of sovereignty in some interesting ways.

It is important to recall that the early space age, which gave birth to ERS technologies, was characterized by fierce competition. For the superpowers, and to a lesser extent for subsequent space powers, large-scale space programs were symbols of national prestige. These autonomous programs were viewed as means of strengthening national security and were seen as necessarily under state control.[10] Military reconnaissance, the direct progenitor of ERS technologies, came to be viewed as a staple in the superpowers' exercise of territorial sovereignty; knowing the adversary's military and industrial capabilities was seen as essential to preventing foreign intervention. Paradoxically, just as the mutual acquisition of nuclear weapons by the superpowers rendered those weapons effectively unusable, the mutual acquisition of satellite reconnaissance technology rendered their territorial space utterly transparent. While satellites may have offered some protection against military intervention, they opened the door to visual intervention.

With the advent of nonmilitary applications of satellite-based remote sensing, concerns about sovereignty and intervention were in some ways heightened. The United States pioneered the use of ERS technologies, launching the moderate-resolution Landsat in 1972; the French-led SPOT (Systeme Probatoire d'Observation de la Terre) began returning higher-resolution data to Earth in 1985; and a Russian consortium entered the market with even higher-resolution images a few years later. Each of these developments raised new sets of issues with respect to sovereignty.

Even today, with international cooperation and the proliferation of commercial and scientific satellite data, the technology's military roots are continually evident. Military agencies still control the lion's share of high-resolution satellite imagery and are reluctant to share it with others. As the militaries of the Cold War superpowers come under pressure to redefine their mission in a post–Cold War era, they have become involved in environmental research.[11] For decades, the security forces of both superpowers did a good deal of inadvertent environmental research, which scientists are now eager to acquire.[12] Yet, in a profound clash of cultures, the highly secretive nature of military technologies and procedures is often at odds with the expectations and conventions of both commercial and research partners—even in the relatively open United States.[13] There is a certain irony in the fact that the Russian consortium, Soyuzkarta, made formerly top-secret, high-resolution imagery from Soviet military satellites available on the mass market while the United States has been reluctant to do so for reasons of national security.[14] At stake is the principle of territorial exclusivity in a world rendered transparent by satellite technology.

This tension reflects the discrepancy between the nonterritorial nature of outer space and the principle of state sovereignty. While the air space above a state's territory falls under that state's jurisdiction, the space above the Earth's atmosphere (outer space) was declared in the 1967 Outer Space Treaty to be a *res communis*, or the common province of humanity.[15] The prohibition of territorial claims in outer space stands in a tense relationship with the efforts of states to enhance their own security through the use of satellites stationed in nonterritorial space. Given the military's leadership role and the resources required to conduct space activities, space issues would seem to reinforce the state-centric model of international relations. Yet the nonterritorial character of space activities poses certain challenges to traditional notions and practices of sovereignty.[16]

Developing Countries, Satellites, and Territorial Sovereignty

Questions of territorial sovereignty were hotly debated during the two sets of negotiations on the use of geosynchronous orbits and on the principles governing the use of satellites in television broadcasting. In both cases, the principle of nonterritoriality prevailed, with implications for environmental remote sensing. In the 1976 Bogota Declaration, the equatorial nations argued that their sovereignty extended 22,300 miles upward to the prized geosynchronous orbits over their territories. They argued that because the existence of the orbit depended upon the Earth's gravity, it was therefore a "physical fact linked" to the planet and thus subject to national sovereignty.[17] However a majority of the UN General Assembly, including most developing countries which preferred a "common heritage of mankind" reading, rejected this interpretation.

The coveted geostationary orbits became part of outer space, although they have, in practice, been more or less governed by a first-come, first-served policy.[18]

In a similar debate over the territorial status of outer space, the former Soviet bloc, and most developing countries held that lack of control over direct broadcasting within their territories would amount to cultural imperialism. The U.S. freedom-of-information perspective, however, which was premised upon the nonterritorial nature of outer space, ultimately prevailed.[19] These two outcomes, each upholding the nonterritorial nature of outer space, entail some important implications for ERS. If states may not claim territorial jurisdiction over the orbits overhead, then claims to sovereignty over the images and data gathered from the satellites stationed in those orbits are substantially weakened.

Beginning in the 1970s, countries without access to satellite technology suspected that an "open skies" policy with respect to ERS might violate their territorial sovereignty. Although they may have harbored such fears earlier with respect to military reconnaissance satellites, the fact that superpower images were not available on the commercial market was a source of comfort. The military secrecy of the superpowers, at least in this case, was an apparent blessing for the Third World. But when NASA espoused an open skies policy with its first launch of the Earth Resources Technology Satellite (later renamed Landsat), some Latin American countries countered that their sovereignty over natural resources extended to the dissemination of information about them. Mexico, for instance, announced that "no data would be collected over Mexican territory from air or space without prior permission."[20]

NASA's response was threefold. First, it argued from international law that there were no legal restrictions on the use of ERS for peaceful purposes. Second, it labeled Landsat an "experimental," rather than an "operational," project until the 1980s. Third, and most effectively, it held out the enticing promise to developing countries that the open dissemination of satellite data would extend, not reduce, their ability to control the development of their resources. In other words, earth-sensing satellites would *amplify* their sovereignty. To add credence to that promise, NASA established an educational program to train scientists from developing countries to use ERS data.[21]

Indeed, many countries came to the counterintuitive conclusion that transparency and the global diffusion of data actually reinforced their territorial sovereignty. By 1980, ten countries had built ground stations and were committed to paying NASA an annual fee of two hundred thousand dollars for data transmission; dozens more were purchasing Landsat images and data tapes. For example, Brazil reported that the first Landsat images resulted in the discovery of several large islands within its territory and a major rectification of Amazon tributaries on its maps. The U.S. Embassy in Mali reported that "the U.S. government has gained a million dollars worth of Malian political mileage" from Landsat.[22]

Satellite data, then, may actually help to prepare a given territory for the exercise of sovereignty. As Thom Kuehls argues in chapter 2, nature is not inherently constituted so as to become subject to state sovereignty, but is rather socially constructed as "territory." Mapping is a crucial element in this social construction. There are no lines of latitude and longitude in nature; overlying the globe with this symbolic organization imposes an artificial order and serves specific political purposes. For this reason, cartography has been labelled "the science of princes."[23] The burgeoning use of geographical information systems (GIS), which use space as a common key between sets of satellite data, can strengthen claims to territorial sovereignty in countries with isolated areas. Thus, the utility of ERS data for mapping and locating resources suggests that the logic of satellites does not always run contrary to the principle of territorial exclusivity.

Nonetheless, developing countries have not always been satisfied with their roles in the emerging ERS regime. By the early 1980s, developing countries were concerned with preserving open and nondiscriminatory distribution of Landsat data, which they felt was threatened by the Reagan administration's proposal to privatize Landsat.[24] Many observers believed that Landsat data should remain a public service, analogous to census, cartographic, and meteorological data, and several studies concluded that Landsat could not be successfully commericialized. Nonetheless, control over Landsat's data was given to EOSAT, a joint venture of Hughes Aircraft and General Electric, under the Land Remote Sensing Commercialization Act of 1984.[25] One of EOSAT's first acts, which was greatly resented by scientists as well as developing countries, was to quadruple the price of each Landsat image.[26]

Cost is just one of the factors limiting the utility of ERS data, from both Landsat and SPOT, in developing countries. Commercial ERS programs tend to be designed with the informational needs of paying customers in mind — primarily multinational agricultural, mining, and oil exploration companies. Because ERS is fundamentally a "high-tech" endeavor based in the industrialized countries, its use elsewhere entails a host of complex technology transfer issues. The use of ERS data requires skills in photogrammetry and computers which are scarce in most developing countries. As a first step in overcoming this problem, the United Nations has initiated Centres for Space Science and Technology Education on a regional basis in developing countries.[27] Yet ERS data continue to remain inaccessible to potential beneficiaries.[28] Nor do the large global change research programs that rely upon ERS data emphasize the kinds of information on land use and ecological change that are most urgently needed in developing countries.[29] The tensions in U.S.-Brazilian environmental research, for instance, have motivated Brazil to launch remote sensing satellites of its own in the 1990s and to broaden its international sources of data.[30]

With the exception of India, China, and Brazil, which have built their own ERS systems, all other developing countries have depended on imagery from Landsat, NOAA, SPOT, and now Russian satellites.[31] But they have not always been satisfied with this arrangement. The United Nations Committee on the Peaceful Uses of Outer Space (COPUOS) has emerged as a champion for the interests of developing countries. In 1991, several developing countries submitted a working paper to the Legal Subcommittee of COPUOS arguing for a kind of affirmative action program, expressed in a new treaty, for developing countries with respect to space technology.[32] They based their position on Article I of the 1966 Outer Space Treaty, which states that the use of outer space "shall be carried out for the benefit and in the interest of all countries, irrespective of their degree of economic or scientific development, and shall be the province of all mankind." Developing countries insisted that, in order for all countries to share equitably in the benefits of space technology, including ERS programs, international cooperation must be based on a system of preferential treatment for developing countries.

They also referred to the Principles on Remote Sensing of the Earth from Outer Space, adopted by the United Nations in 1986, which call for international cooperation in the use of ERS technologies.[33] Thus, taking a notably different tack from that of the equatorial states in the earlier debate on geosynchronous orbits, developing countries cite the common property status of outer space as a rationale for preferential treatment in the emerging ERS regime.

The United States and the United Kingdom, however, countered the developing countries on the basis of traditional sovereignty claims. Attempting to appease them, on the one hand, they cited many examples of bilateral and multilateral programs in space technology. On the other hand, however, they claimed that any attempt to impose legal obligations for cooperation would undercut states' sovereign right to decide the sorts of joint programs in which they would participate. The developing countries split ranks when Brazil and Nigeria concurred with this line of argument, scuttling efforts to negotiate a new treaty. Their effort to promote preferential treatment may have been successful, though, in serving notice to the space powers that they should work harder to ensure that ERS and other forms of space technology are beneficial to developing countries.

Developing countries have apparently embraced ERS, but not without some reservations about the technology's impact on their territorial sovereignty. One feature of ERS is its dual role in providing an information base and as a technology for monitoring. Sovereignty has traditionally been invoked to shield states from external intrusion, yet satellites render territory effectively naked.[34] While the transparency afforded by satellite observations can aid in the monitoring of international law, it can also be interpreted as a tool for foreign intervention. Compliance with international environmental agreements, for instance, has tended to be voluntary, with nongovernmental organizations frequently functioning in a watchdog capacity. When mandated, verification of compliance has generally proceeded through self-reporting. Thus, certain developing countries have expressed the concern that ERS could foster "green conditionality" and other types of "eco-imperialism." Just as satellites can be used to monitor treaty compliance, so too can they be used for industrial espionage. Many observers believe that "in the future, commercial remote sensors will not only be able to detect pollutants leaving a factory, but determine what a factory is producing."[36] For these reasons, developing countries insisted that the Earth Summit documents adopted at Rio de Janiero in 1992 contain no references to the use of ERS for "monitoring," but only for "observation."[36] Whether the semantical distinction between surveillance and observation will translate into practice remains to be seen.

Commercial Satellites and Reconfigured Sovereignty

Ironically, it has been primarily the United States, not the developing countries, which has sought to place restrictions on ERS data and technology in the name of territorial sovereignty. The U.S. restrictions harken back to the technology's roots in military reconnaissance. In 1978, President Carter upheld the Pentagon's interests over NASA's by signing a presidential directive that set ten meters as the resolution limit for, nonmilitary remote sensing.[37] But the entry into the international market of SPOT, with a resolution of ten meters, and Soviet satellite photographs of roughly five-meter resolution, soon made this rule essentially obsolete. The Reagan administration deleted the rule in 1988 after being persuaded that it put American satellite operators, especially the now-privatized Landsat data marketing firm, at a distinct disadvantage. In an effort to uphold traditional national security interests, the new directive granted veto power to the secretaries of defense and state over the licensing of U.S. commercial remote sensing satellites.[38] But U.S. officials were at a loss to describe how they would enforce a ban on the dissemination of pictures from space, since the United States no longer enjoyed a monopoly on Earth-scanning satellites—even in the West. And, of course, the most likely beneficiaries of American regulations would be foreign satellite operators. As a SPOT spokesperson observed, "Open skies, open access is a precondition of commercial success in the remote-sensing industry."[39] The Clinton administration went beyond the Reagan rule with its commitment to consider favorably licensing applications for ERS systems whose capabilities are already available or are in the planning stages.[40]

The emergence of high-resolution satellite imagery on the world market provides an interesting example of how the practices of sovereignty can be driven by technological developments and globalization. It makes little sense to place domestic restrictions on high-resolution data which are easily accessible from foreign suppliers. A technology that cannot be controlled by a single government is impossible to contain; satellite images can only be suppressed if the data are sent to a ground station under the control of the censoring government. In a classic sovereignty bargain, the United States has been forced to revise its conceptions of national security in order to promote its own industrial competitiveness.

It might be an overstatement to declare, as some have, that satellites have "abolished the concept of distance,"[41] but it is certainly the case that the practices associated with territorial sovereignty are being revised. There is no single, straightforward logic to ERS technology. Certainly, it still bears the imprint of its origins in military reconnaissance, the root purpose of which was to protect the superpowers' territorial integrity. Moreover, ERS is being used by some developing countries to expand and reinforce their claims to sovereignty within their borders. Yet the emergence of ERS data on the world market has dramatically eroded the ability of states to control information about the resources within their borders. The almost universal availability of ERS data has rendered much of the world transparent; its global nature appears to be undercutting the characteristically modern conceptualization of Earth as territorially demarcated. If, as David Harvey suggests, modernity located "the other" in a specific place "in a spatial order that was ethnocentrically conceived to have homogenous and absolute qualities,"[42] then ERS, by virtue of its globality and its transparency, challenges this spatial order, and thus stands at the cusp of the modern and the postmodern.

Sovereignty Bargains

Just as it would be an oversimplification to say that transboundary environmental degradation necessarily subverts state sovereignty, so too would it be simplistic to say that international cooperation axiomatically subverts state sovereignty. Paradoxically, in an interdependent world, cooperation may help to sustain the institution of sovereignty. Sovereignty may be disaggregated, with autonomy, control, or authority in one area being traded for greater autonomy, control, or authority in another area. While such sovereignty bargains may not destroy sovereignty, they do alter it. While there is nothing new in the fact that states often make such trade-offs, the contemporary logic of ERS seems to be pushing states to sacrifice autonomy in exchange for greater collective problem-solving capacity.

The multiple-use character of ERS data has compelled states to try to strike a balance among competing military, economic, and environmental interests. As a corollary, international sovereignty bargains manifest domestically as bureaucratic competition among state agencies. The huge economies of scale to be gained through pooling ERS resources have encouraged states to sacrifice some autonomy and control in exchange for better access to information. The end of the Cold War has made a whole range of new sovereignty bargains feasible.

Consider the incremental relaxation of restrictions on ERS technology by the U.S. government since the late 1970s. The apparent contradiction between U.S. industrial competitiveness and perceived military interests is gradually being resolved in favor of the former, an uneasy settlement driven by the globalization of technological change. Whereas resolutions finer than ten meters were once deemed a serious security risk, even domestically, the United States is now licensing commercial ERS systems with a resolution of only one meter.[43] This does not mean that security is no longer a top concern for states, but it does suggest that the meaning of security, like that of sovereignty, is being revised. Nonetheless, traditional security arguments have also been used to promote industrial competitiveness. For instance, proponents of easing restrictions on high-resolution ERS technology argued that, "Failure to allow the (U.S.) remote sensing industry to grow aggressively will only encourage the development of suppliers that may be impossible to control in a time of crisis."[44] Apparently, a parallel debate has been occurring in Russia.[45]

The relationship of military agencies to commercial and scientific ERS has always been a tense one.[46] Yet the military does a good deal of inadvertent earth science. Because military and civilian ERS systems often duplicate each other's work, a merger of some programs would be more efficient. National security concerns, however, may pose a critical obstacle.[47] Even though the U.S. military has been actively promoting access to formerly classified data and facilities in order to justify its post–Cold War budget, cultural barriers often prevent researchers from obtaining usable military data. Again, global technological change provides the impetus for new sovereignty bargains. Consider, for instance, the fact that once Europe's Earth Resources Satellite-1 (ERS-1) began returning the same kind of gravity data as the U.S. Navy's Geosat system, the U.S. military suddenly became more responsive to civilian researchers.[48]

In the long run, the most important types of ERS-related sovereignty bargains are likely to be those embodied in international cooperative endeavors. At the most general level, the Outer Space Treaty hinges upon a sovereignty bargain in which states accept responsibility for all activities of their nationals in outer space in exchange for the recognition of their right to use outer space for peaceful purposes.[49] On a more instrumental level, the efficiency of cooperative programs provides a powerful incentive for states to collaborate on a wide range of space programs, including ERS.[50]

Before the end of the Cold War, there was a political consensus on both sides of the Iron Curtain linking space to national security objectives — not only for projects with an obvious military value, but also for civilian prestige projects. Since the end of the Cold War, the alliance of space with narrowly defined national interests has deteriorated, as the proliferation of U.S.-Russian space programs demonstrates.[51] That development, combined with a general mood of fiscal conservatism, has sparked a major increase in the number and scope of cooperative space programs. In the words of one observer, "Now that the Cold War is over, we can afford to be efficient."[52] Sovereignty has not been "eroded," but the range of acceptable sovereignty bargains on space projects has dramatically increased with the end of the Cold War, prompting the proliferation of international ERS programs.

Just as important for international cooperation in ERS have been the new exigencies of global environmental research. A patchwork of transnational scientific research programs has sprung up in the last decade, including: the Man and the Biosphere Programme, the International Biosphere-Geosphere Programme, and the World Climate Research Programme. To a very great extent, these programs, spearheaded by international organizations and nongovernmental scientific organizations, rely upon satellite data provided through national space agencies. These international programs seek to achieve a "worldwide synergy of local research" by bringing together the financial and organizational capabilities of governments with the intellectual capacity of the world's scientific community.[53]

Though NASA is undoubtedly the major player in most of these programs, virtually every ERS project has an international component. Most of the satellites launched under NASA's Mission to Planet Earth program have carried instruments from other countries, or else have transmitted data to other countries. Likewise, Japan's new Advanced Earth Observing Satellite (ADEOS) carries two U.S. and one French instruments.[54] The principle international coordinating body for Earth observations is the Committee on Earth Observations Satellites (CEOS), which was created in 1984 in connection with the annual G-7 Economic Summit, and whose membership includes all national and supranational space agencies. A smaller body, the Earth Observations International Coordination Working Group (EO-ICWG), provides a more restricted forum for Canada, Europe, Japan, and the United States to plan the International Earth Observing System (IEOS) for the 1990s and beyond.[55]

The kinds of voluntary arrangements represented by CEOS and EO-ICWG are emblematic of a particular kind of sovereignty bargain whereby states sacrifice some degree of autonomy and control over technological and informational resources in exchange for the benefits of collaboration, which include cost savings and intellectual synergy. But these bargains are not without their drawbacks. Once states become dependent upon a continued supply of Earth observation data that they do not themselves control, they are faced with the dilemma that their access to data is perpetually at the mercy of other states' budgetary processes. For instance, while the European Space Agency (ESA), having been once burned by NASA in the Spacelab project, insisted upon effective sovereignty over the elements it contributed to the space station, it nonetheless remains hostage to NASA's budgetary roller coaster. According to a clause in the 1988 ESA-NASA agreement, which is standard fare in joint international space ventures, states' obligations are "subject to availability of funds."[56] Given that NASA's Earth Observation System (EOS) program has already been scaled back twice, there is a strong likelihood that budgetary politics will interfere with other cooperative ERS endeavors.

Various proposals have been introduced for an international regime that would promote the efficient and systematic use of ERS technologies.

One proposal, initiated by the Society of Japanese Aerospace Companies, is for a World Environment and Disaster Satellite Observation System (WEDOS) that would monitor natural and man-made disasters on all time scales.[57] A more comprehensive proposal, under discussion since the mid-1980s, is for ENVIROSAT, a regime analogous to INTELSAT and INMARSAT, to provide climate, meteorological, ocean, and land observations. Regime members would contribute to the capital expenses of the system by paying in proportion to use, and users would pay commercial fees for services.[58] This sort of sovereignty bargain would simultaneously increase states' mutual dependence on ERS technology and data, while making it more difficult for states to renege on prior commitments for budgetary reasons. In spite of the obvious benefits, an international ERS regime, whose users would include government agencies, academic researchers, multinational corporations, local communities, and nongovernmental orgranizations, would be far more difficult to negotiate than a regime like INTELSAT, which serves only the communications industry. At stake would be critical questions about the ownership of knowledge.

The sorts of sovereignty bargains embodied in international cooperative arrangements involving ERS technology do not necessarily signify "the erosion of sovereignty." Rather, they represent concrete choices by states to sacrifice some elements of authority, autonomy, and control for others. Those choices, which are never free choices but are constrained and largely driven by the dynamics of technological globalization, engender the reconfiguration of political space and the renegotiation of sovereignty.

The View from Space: Sovereign Knowledge?

Because of its central role in the dissemination of knowledge about the Earth, the most interesting political questions involving ERS technology pertain to the control of knowledge and information and the purposes to which they are applied, bringing us back to the neglected informational dimension of sovereignty.[59] Knowledge and sovereignty are conceptual kin; both sorts of claims are fundamentally about whose voice is to be regarded as authoritative. And because the quest for scientific knowledge is a cornerstone of modernity, these issues inevitably return us to the larger question of the relationship of ERS to modernity.

Information has inherent public-goods attributes, so that governments are likely to continue to play a significant role in ERS funding and application.[60] And although information is "slippery" by nature, its production and dissemination are costly and its close relationship to power can kindle conflicts over its control and possession. Consider the following disputes over access to satellite-derived information. Developing countries' lack of confidence in an uninterrupted supply of ERS data from the United States, particularly after the privatization of Landsat, prompted the largest of them to build their own remote sensing satellites.[61] Researchers harbored similar sentiments, but they lack the option of building their own satellites. According to one scientist, the tenfold increase over the 1970s price in the cost of Landsat data after privatization effectively impeded a good deal of scientific research.[62] Both government agencies and scientific researchers feel that commercialization threatens their access to data. SPOT, for instance, implemented a policy in 1989 of giving preferential service to its largest customers, the oil and mining industries, potentially placing certain government agencies at a disadvantage in obtaining urgently needed data.[63] More recently, European governments have threatened to launch a "data war" by attempting to restrict commercial access to ERS data from weather satellites. Their moves have inflamed researchers, who claim that scientific and commercial data are not easily distinguishable.[64] In a similar vein, ensuring data consistency is a central concern for researchers, whereas commercial competitiveness entails exactly the opposite: capabilities, image size, and hardware are differentiated as much as possible to prevent commercial users from switching systems.[65] In June 1995, the World Meteorological Organization voted to restrict the availability of some kinds of weather data, "in effect creating a new commodity which can be encrypted, bought and sold, licensed, and controlled in a way that such data had not been before."[66] All of these points of dissension have implications for issues of control and authority in an information age, issues that include but are not limited to state sovereignty.

New technologies do not emerge as neutral tools; rather, they arise in a context of ongoing struggles for control and authority, amplifying certain voices and inhibiting others. Any technology as useful as ERS inevitably becomes the object of great contention. Is there, then, a distinctive logic to ERS as an artifact/idea that tends to legitimize or empower certain voices over others? If knowledge and information are preeminent sources of power in late modernity, then what do the globality, the transparency, and the high-tech nature of ERS entail for the distribution of social and political power?

At first glance, the logic underlying ERS appears to be profoundly technocratic. The skills required to operate satellites and sensors, and to decipher ERS data and, imagery, are concentrated in an elite group of technicians and scientists from industrialized countries.[67] At times, ERS experts exhibit an almost missionary zeal reminiscent of the Baconian technocratic ideal. Space technology is said to offer "unlimited perspectives on ourselves, the world, and the cosmos around us."[68] One champion of ERS technology, Thomas Becker, even suggests that human survival depends upon it: "The great opportunity for progress in the world in the 20th century was physics, which built the world we live in. The great opportunity for creative progress in the next century will be Earth Science. It will determine if humankind is in the universe to stay."[69] As the capabilities of ERS technology expand, such sentiments may become even more prevalent. Already, a marine biologist can sit at his computer and "get information from a free-ranging whale anywhere on Earth."[70]

While the technocratic potential of ERS appears to be evident, other forces could compel the architects of ERS technology to become more accountable to its users. The very multiplicity of ERS users — ranging from research scientists in many fields to the extractive industries to environmentalists — suggests the potential for a diffusion of power along multiple channels. The state may be an important channel, but it is neither the only one nor is it a univocal one. As "Big Science" projects lose their appeal in a time of budgetary conservatism, and as their prestige value is diminished with the end of the Cold War, space agencies must increasingly justify ERS programs in terms of their users' requirements. One space scientist calls this a "thoroughly postmodern approach," stating that, "No longer will the development of new technology be driven by an elite of scientists and engineers, but a broader base of consultation will be required with the many user constituencies."[71]

ERS is a multifaceted artifact/idea incorporating sometimes contradictory tendencies. On the one hand, the global view afforded from the vantage point of space seems especially conducive to notions of "planetary management" and the centralization of power. Indeed, in the discourse surrounding ERS, terms like "managing the planet" and "global management" abound.[72] Yet global science is inherently decentralized, depending upon "countless loosely knit and continually shifting networks of individual researchers—most of whom resist outside intervention—in communication that crisscrosses the borders of well over a hundred sovereign nations."[73] The decentralized nature of global science is likely to have important social and political implications for efforts to cope with ecological interdependence, implications which are beyond the scope of this chapter.

While the global science based upon ERS data has many of the earmarks of a mammoth technocratic enterprise, it is not immune to public opinion; nor are its fruits available only to the elite. NASA's Mission to Planet Earth program, for instance, was conceived as a vehicle for restoring the confidence of Americans, newly concerned about the environment, in the space agency after the Challenger disaster.[74] Even in Japan, popular environmental concern shifted the emphasis of its new Earth resources spacecraft, ADEOS, away from pure research objectives.[75] In the future, ERS satellite systems could provide citizen groups with the means to verify compliance not only with, environmental treaties, but with arms control treaties as well, with potentially interesting ramifications for the tension between state sovereignty and popular sovereignty.[76]

ERS data can facilitate the localization of control in some surprising ways. Perhaps most interesting is the use of satellite data by indigenous peoples for mapping their customary land rights and documenting the role of the state and multinational corporations in environmental destruction. Environmental advocacy groups and indigenous rights groups in Indonesia, Nepal, Thailand, and the Pacific Northwest are using satellite-generated data to reterritorialize their political practices to an extent previously inconceivable.[77] While ERS data may deterritorialize political practice at the level of the nation-state, it seems to be having exactly the opposite effect at the local level. Thus, ERS technologies may facilitate challenges to state sovereignty from below, a promising development for groups trying to assert local control.

Because the use of ERS data in developing countries raises a host of complex cultural, political, and ethical issues, not all observers see this sort of technology transfer in a positive light. For instance, Masahide Kato is critical of nonprofit groups based in industrialized countries who supply satellite-generated information to remote areas of developing countries. He believes they are representives of a "globalist technosubjectivity" which renders the territories of indigenous peoples as resources.[78] Indeed, satellites seem to offer the tantalizing prospect of "sovereign knowledge," or knowledge with supreme authority. As one early enthusiast proclaims, they "show vast terrains in correct perspective, from one viewpoint, and at one moment in time."[79] But, that "one viewpoint" is generally located in the North and that "one moment in time" cannot capture centuries of past environmental abuse, a fact that may prove profoundly disadvantageous for developing countries when ERS data are used to assign responsibility for ecological degradation.[80]

While Kato perhaps too quickly condemns ERS technology, which we have seen can be used to promote the interests of indigenous peoples, his critique reveals two interrelated issues of political culture implicit in ERS as an artifact/idea: the control of knowledge (who controls it and for what purposes) and the constitution of knowledge (what counts as knowledge). By employing ERS data, environmental and indigenous rights groups demonstrate that it can be translated into usable knowledge for purposes of cultural and ecological preservation, but they simultaneously legitimize it as a source of credible knowledge. The mantle of scientific objectivity that is gained when traditional practices are transposed into the language of GIS is not cost-free. As one scientist cautions,

[While] spatial information technology may enable local people to make claims against the state, this power comes with a price—it destroys the fluid and flexible nature of their traditional perimeters.... While maps can be an empowering tool, helping a local community define itself in relationship to the landscape and to the political forces that shape and influence it, maps can also be used to disinherit them.[81]

Moreover, the voyeuristic nature of photography, including satellite imagery, may promote a view of nature that is antithetical to the long-term ecological goals of grass roots groups. In her famous essay, *On Photography*, Susan Sontag argued that, "cameras implement the instrumental view of reality. [They] arm vision in the service of power—of the state, of industry, of science."[82] Some grass roots groups are wagering that ERS can also "arm vision in the service of power" at the local level, and thus serve the cause of resistance. Their efforts, however, are too recent to draw any conclusions at this point.

Users of satellite-generated Earth data have powerful cultural and rhetorical tools on their side—specifically Enlightenment ideals about the liberating power of knowledge. According to one viewpoint programs employing ERS information should be based on the premise that "greater knowledge leads to greater wisdom,"[83] a premise that deserves scrutiny. If the link between knowledge and wisdom is weak, the link between knowledge and power is more palpable. Indeed, a core assumption of the architects of ERS systems is that they offer "a whole new tool with which to understand our own world, and once we understand it, we can manage it."[84] Such statements seem to presuppose a specifically modern conception of agency and responsiblity, with a rational autonomous self capable of knowing (and thereby controlling) the Other embodied in the natural "environment." This is exactly the form of Western subjectivity which Franke Wilmer critiques in chapter 3 of this book. If this hallmark of modernity is actually at the root of the global environmental crisis, then the faith in ERS technology may be fundamentally misplaced.

Given the deep entrenchment of the knowledge/power nexus as a cultural cornerstone of modernity, to question the need for more information approaches heresy. Yet, given the stakes, we must ask whether ERS data will tell us what we need to learn. NASA's Earth Observing System (EOS) will produce an unprecedented quantity of data, at a cost of perhaps twenty billion dollars; its data information system (EOSDIS) will be the largest data-handling system ever constructed, with a capacity of fourteen petabytes (a petabyte is 10^{15} bytes).[85] A global ERS system is expected to provide "the long-term measurements to determine the habitability of the Earth" and guide policy makers in addressing global environmental change.[86] With less than 5 percent of Landsat's data having ever been used,[87] it is quite possible that ERS technologies will generate more information overload than usable knowledge. According to the World Meteorological Organization, a satellite-based Global Climate Observing System, with EOS as its core, "will require substantial resources, but the costs to society from continuing the present level of uncertainty about climate change are very much larger."[88] What those costs are, who bears them, and how ERS data will decrease them, are not discussed.

Will the knowledge gained through ERS technology tell us how to live sustainably? The answer to the question will depend largely upon who uses the information and to what purposes it is applied.

Conclusion

The transparency, globality, and technological sophistication associated with ERS technologies entails multiple, and sometimes contradictory, implications for the institution of sovereignty. As we have seen, there is no single political logic for ERS as an artifact/idea, which is partly because technologies tend not to be monolithic cultural constructs. But, more importantly, the cross-cutting logic stems, in many ways, from the fact that ERS stands at the intersection between modernity and post-modernity. First, ERS is fundamentally an information age technology, with all the concomitant implications for the power/knowledge nexus that arise when data make up the currency of power and materialist conceptions of power are not fully applicable.

Second, ERS contributes to the unbundling, but not the abolition, of territoriality. While the transparency and globality associated with ERS technologies very often deterritorialize state practices, they are also capable of bolstering territorial sovereignty for developing countries with remote regions. More surprising, however, are the ways in which they are helping local environmental and indigenous groups to reterritorialize their political practices.

Third, while ERS programs have their roots in the balance-of-power politics characteristic of the modern nation-state system, since the end of the Cold War they tend to exemplify the sorts of sovereignty bargains required by scientific and environmental (and, to a lesser extent, commercial) cooperation. If modernity is interpreted as the enclosure of the globe via the twin institutions of state sovereignty and private property, then ERS technologies at once epitomize and challenge that trend. On the one hand, by making visible the invisible, satellite imagery renders nature subject to claims of ownership and control. On the other hand, in light of the globality and transparency inherent in ERS technologies and the emphasis on environmental cooperation, ERS has the potential to become a tool in the revisioning of nature as a global commons. Indeed, this is the thrust of much of the discourse surrounding environmental ERS.

Finally, there is the tension between the universal, totalizing perspective of the sovereign gaze, and the application of ERS technologies to popular sovereignty through the decentralization of scientific and political control. According to John Ruggie, the transformative potential of global ecology as a basis for a postmodern social episteme lies in "the underlying structural premise of ecology [of] holism and mutual dependence of parts." He also suggests that the study of the emergence of "multiperspectival institutional forms" is key to understanding the transition to postmodernity.[90] By itself, holism could legitimate and reinforce the uniperspectival attitude of the sovereign gaze. While this distorted interpretation of ecological principles is at times evident among proponents of ERS programs, it is countered by a contrary tendency toward pluralism and the proliferation of voices among users of ERS technologies and data. Thus, the relationship of ERS to sovereignty can be problematized and analyzed in terms of how, as an artifact/idea, it operates at the crossroads between modernity and postmodernity.

Notes

1. I am grateful to Ronald Mitchell, Lisa Schafer, Janice Thompson, and Mark Zacher for their helpful comments on earlier versions of this paper.
2. Langdon Winner, "Artifact/Ideas and Political Culture," in Albert H. Teich, ed., *Technology and the Future*, 6th ed. (New York: St. Martin's Press, 1993): 283–92.
3. Committee on Earth Observation Satellites (CEOS), "The Relevance of Satellite Missions to the Study of the Global Environment," produced for the UNCED conference at Rio de Janeiro (Washington, D.C.: CEOS, 1992). On the multitude of uses for ERS, see Doug Stewart, "Eyes in Orbit Keep Tabs on the World in Unexpected Ways," *Smithsonian* 19 (December 1988): 70–76.
4. John H. McElroy, "INTELSAT, INMARSAT, and CEOS: Is ENVIROSAT Next?" in Gordon MacDonald, ed., *Proceedings of the Conference on*

Space-Based Monitoring of the Global Environment (La Jolla, Calif.: Institute on Global Conflict and Cooperation, 1992).

5. Anthony Giddens, *The Nation-State and Violence* (Berkeley and Los Angeles: University of California Press, 1987), 52, 309; Michel Foucault, *Discipline and Punish: The Birth of the Prison* (New York: Vintage, 1979).

6. Joseph N. Pelton, "Organizing Large-Scale Space Activities: Why the Private Sector Model Usually Wins," *Space Policy* 8 no. 3 (August 1992): 234.

7. "A Problem as Big as a Planet," *Economist* (5 November 1994): 83–85.

8. Michel Foucault, *Power/Knowledge: Selected Interviews and Other Writings*, C. Gordon, ed., and trans. C. Gordon, L. Marshall, J. Mepham, and K. Soper (New York: Pantheon, 1980); Mark Poster, *Foucault, Marxism, and History: Mode of Production vs. Mode of Information* (Cambridge: Polity, 1984); Frederic Jameson, "Postmodernism, or the Cultural Logic of Late Capitalism," *New Left Review* 146 (July-August 1984): 53–92.

9. John Gerard Ruggie, "Territoriality and Beyond: Problematizing Modernity in International Relations," *International Organization* 47, no. 1 (Winter 1993): 171.

10. Pelton, "Organizing Large-Scale Space Activities," 242.

11. While military agencies may be struggling to redefine their mission, the end of the Cold War does not spell the obsolescence of military reconnaissance. See Jeffrey T. Richelson, "The Future of Space Reconnaissance," *Scientific American* 264, no. 1 (January 1991): 38–44; and John Trux, "Desert Storm: A Space Age War," *New Scientist* 27 (July 1991), pp. 30–34. On NASA's Mission to Planet Earth program, and the conversion of missiles to environmental purposes, see Ann Florini and William Potter, "Goodwill Missions for Castoff Missiles," *Bulletin of the Atomic Scientists* (November 1990): 25–31.

12. Robert Dreyfuss, "Spying on the Environment," *Earth Action* 6, no. 1 (February 1995): 28–36.

13. On the disadvantages of involving the military in ERS, see Ronald Diebert, "Out of Focus: "U.S. Military Satellites to the Environmental Rescue," in Daniel Deudney and Richard Matthews, eds., *Contested Grounds: Security and Conflict in the New Environmental Politics* (Albany: State University of New York Press, forthcoming).

14. Leonard S. Spector, "Keep the Skies Open," *Bulletin of the Atomic Scientists* (September 1989): 16. A 1988 image, apparently bought from Soyuzkarta, of an air base in Nevada which the Pentagon does not admit exists, has turned up in instructions for a Testor model airplane and on the cover of *Popular Science*. See "Get Satellite Imagery Policy in Focus," *Aviation Week and Space Technology* (February 21, 1994): 124.

15. "Legislating the 'Last Frontier,'" *UN Chronicle* 29 (December 1992): 54; "Treaty on Principles Governing the Activities of States in the Exploration and Use of Outer Space, including the Moon and other Celestial Bodies," *United Nations Treaty Series* 610: 205 (1967). The question of remote

sensing satellites and territorial sovereignty had been partially resolved when President Kennedy promised the world free access to weather data in exchange for an implicit recognition of outer space as a global commons.

16. David Green, "The Reassertion of Social Aspects of Science and Technology," *Space Policy* 10, no. 3 (August 1994): 242.

17. Carl Q. Christol, *The Modern International Law of Outer Space* (Oxford: Pergamon, 1982): 468, quoted in Joanne Irene Gabrynowicz, "Bringing Space Policy into the Information Age," *Space Policy* 8, no. 2 (May 1992): 168.

18. See Stephen A. Doyle, "Space Law and the Geostationary Orbit: The ITU's WARC-ORB–85088 Concluded," *Journal of Space Law* 17 (1989): 13–21; Report of the Committee on Peaceful Uses of Outer Space, UN Doc. A/49/20 (1994).

19. M.J. Peterson argues that the nonterritorial perspective won out not just for reasons of power and interests, but because the analogical reasoning linking outer space to the high seas and Antarctica had become so firmly entrenched. See "Extending International Law to New Fields of Endeavor: Developing Outer Space Law," unpublished manuscript, 1995. See also "Legislating the 'Last Frontier.'"

20. Quoted in Pamela Mack, *Viewing the Earth: The Social Construction of the Landsat Satellite System* (Cambridge, Mass.: MIT Press, 1990): 187.

21. David T. Lindgren, "Commercial Satellites Open Skies," *Bulletin of the Atomic Scientists* (April 1988): 34.

22. Mack, *Viewing the Earth*, 189–92.

23. J.B. Harley, "Maps, Knowledge, and Power," in Denis Cosgrove and Stephen Daniels, eds., *The Iconography of Landscape: Essays on the Symbolic Representation, Design, and Use of Past Environments* (Cambridge: Cambridge University Press, 1988): 227–312.

24. Mack, *Viewing the Earth*, 188.

25. "Report Criticizes Landsat Commercialization," *Aviation Week and Space Technology* 118 (May 9, 1983): 18; Mack, ibid., 206.

26. Eliot Marshall, "Landsats: Drifting toward Oblivion?" *Science* 243 (February 24, 1989): 24.

27. Adgun Ade Abiodun, "An International Remote Sensing System," *Space Policy* 9, no. 3 (August 1993): 183.

28. For instance, the public health sector, which could employ ERS technology to combat many diseases, had yet to become a user as of 1989. Peter Jovanovic, "Satellite Medicine: Space Technology in Primary Health Care," *World Health* (January 1989): 18.

29. Committee on Earth Studies, Space Studies Board, National Research Council, *Earth Observations from Space: History, Promise, and Reality* (Washington, D.C.: National Academy Press, 1994): II–5.

30. Frederic Golden, "A Catbird's Seat on Amazon Destruction," *Science* 246, no. 4927 (October 13, 1989): 201.

31. "Chinese Developing Satellites for Earth Resources Exploration," *Aviation Week and Space Technology* (July 22, 1985): 81–84.

32. UN document A/AC.105/C.2/L.182, April 9, 1991, submitted by Argentina, Brazil, China, Mexico, Nigeria, Pakistan, the Philippines, Uruguay, and Venezuela.

34. N. Jasentuliyana, "Ensuring Equal Access to the Benefits of Space Technologies for All Countries," *Space Policy* 10, no. 1 (February 1994): 11.

34. Molly K. Macauley, "Collective Goods and National Sovereignty: Conflicting Values in Global Information Acquisition," in *Proceedings of the Conference on Space Monitoring of Global Change* (San Diego: Institute on Global Conflict and Cooperation, 1992): 31–55.

35. Rheem, 16.

36. Dreyfuss, "Spying on the Environment," 36.

37. Peter D. Zimmerman, "Photos from Space: Why Restrictions Won't Work," *Technology Review* 91 (May-June 1988): 48. The resolution of Landsat images at the time was thirty meters, while military reconnaissance satellites had a resolution of less than half a meter. See Nicholas Daniloff, "How We Spy on the Russians," *World Politics Magazine* (December 9, 1979): 24–34.

38. Theresa M. Foley, "Pentagon, State Department Granted Veto over U.S. Remote Sensing Satellites," *Aviation Week and Space Technology* (July 20, 1987): 20–21. In an interesting twist, the Pentagon became a paying customer of SPOT for images of Soviet military installations; while it had millions of its own images, these were classified and so could not be published in its reports. See William M. Arkin, "Long on Data, Short on Intelligence," *Bulletin of the Atomic Scientists* 43, no. 5 (June 1987): 5.

39. Daniel Charles, "U.S. Draws a Veil over 'Open Skies,'" *New Scientist* 116, no. 1585 (November 5, 1987): 29. Many space experts have argued that commercial ERS can promote peace by lessening military secrets and by promoting the independent verification of arms control treaties. See William J. Broad, "Private Cameras in Space Stir U.S. Security Fears," *New York Times* (August 25, 1987): Michael Krepon, ed., *Commercial Observation Satellites and International Security* (New York: St. Martin's, 1990).

40. Office of the Press Secretary, the White House, "Foreign Access to Remote Sensing Capabilities," *Space Policy* 10, no. 3 (August 1994): 243–44.

41. Kiran Karnik, "Remote Sensing: The Indian Experience," *UNESCO Courier* 46 (January 1993): 17.

42. David Harvey, *The Condition of Postmodernity: An Enquiry into the Origins of Cultural Change* (Cambridge: Blackwell, 1989): 252.

43. The Clinton adminstration's policy is outlined in the White House, "Foreign Access to Remote Sensing Space Capabilities." The controversy preceding this policy is discussed in John Morrocco, "Lawmakers Warn Clinton on

Satellite Imagery Sales," *Aviation Week and Space Technology* (November 22, 1993): 38.

44. "Get Satellite Imagery Policy in Focus."

45. James Asker, "High-Resolution Imagery Seen as Threat, Opportunity," *Aviation Week and Space Technology* (May 23, 1994): 51–53.

46. Interestingly, the political roots of NASA's earth science programs lie partly in the Reagan administration's desire to dispel the popular impression that it was only interested in space research for its military applications. "NASA Floats a Global Plan," *Science* 217 (September 3, 1982): 916.

47. Committee on Earth Studies, *Earth Observations from Space*, ix, 9–10.

48. Richard Kerr, "The Defense Department Declassifies the Earth—Slowly," *Science* 263 (February 4, 1994): 625–26.

49. In a sense, this can be viewed as a consolidation of sovereignty. Just as states became responsible for their nationals at sea, thereby effectively eliminating mercenarism and piracy, states are now responsible for their citizens' activities in outer space. See Janice Thomson, *Mercenaries, Pirates, and Sovereigns: State-Building and Extraterritorial Violence in Early Modern Europe* (Princeton, N.J.: Princeton University Press, 1994).

50. Thus, the market for launching satellites is a global one for the simple reason that demand in any one country is insufficient to support that country's launch capacity. See Jack Scarborough, "Free Trade and the Commercial Launch Industry," *Space Policy* 8, no. 2 (May 1992): 109.

51. The unique relationship of the United States to its NATO allies during the Cold War made certain arrangements acceptable that would be virtually unthinkable today. Europe, for instance, agreed in the 1970s to give full "jurisdiction and control" of its Spacelab to the commander of NASA's space shuttle. A recent ESA-NASA agreement, however, gives Europe full sovereignty over the elements of the space station that it contributes. See George van Reeth and Kevin Madders, "Reflections on the Quest for International Cooperation," *Space Policy* 8, no. 3 (August 1992): 221–22.

52. Pelton, "Organizing Large-Scale Space Activities," 244.

53. Rene Lefort, "A Worldwide Synergy," *UNESCO Courier* 45 (July–August 1992): 42; Eugene Bierly, "The World Climate Program: Collaboration and Communication on a Global Scale," *Annals of the American Academy Political and Social Science* 495 (January 1988): 110.

54. Hatoyama-Machi, "Japanese Earth Satellites Spawn Multiple User Groups," *Aviation Week and Space Technology* 133 (August 13, 1990): 70–72.

55. Committee on Earth Studies, *Earth Observations from Space*, 27–31.

56. Van Reeth and Madders, "Reflections on the Quest, "226.

57. Joan Johnson-Freese, "Development of a Global EDOS: Political Support and Constraints," *Space Policy* 10, no. 1 (February 1994): 45–55. One of

the strongest opponents of the Japanese proposal has been NASA, which views EDOS as a potential competitor with its own EOS program.

58. John L. McLucas and Paul M. Maughan, "The Case for ENVIROSAT," *Space Policy* 4, no. 3 (August 1988): 229–39; Stephen Day, "Is the Next Step, InRemSat?" *Satellite Communications* 15, no. 4 (April 1991): 22; Abiodun, "An International Remote Sensing System: A Possibility," 179–84.

59. *Data* is not the same thing as *knowledge*. Data are generated from simple empirical observation, while knowledge entails causal understandings and typically involves some degree of generalizability. Creating knowledge out of data requires analytical abilities that go beyond the observational capacities of instruments such as those found on earth-orbiting satellites.

60. Molly K. Macauley and Michael A. Toman, "Supplying Earth-Observation Data from Space," *Space Policy* 8, no. 1 (February 1992): 17.

61. Frederic Golden, "A Catbird's Seat on Amazon Destruction: Brazil's Space Agency Is Playing. an Expanded Role in Monitoring the Nation's Environment," *Science* 246, no. 4927 (October 13, 1989): 201.

62. Christopher Fotos, "Commercial Remote Sensing Satellites Generate Debate, Foreign Competition," *Aviation Week and Space Technology* 129 (December 19, 1988): 49–50.

63. Spector, "Keep the Skies Open," 16; Group of Experts, UNISPACE 82 Conference, "International Remote Sensing System Proposed by Experts," *UN Chronicle* 22, no. 2 (February 1985): 20.

64. Andrew Lawler, "U.S., Europe Clash over Plan to Set Policy on Data Access," *Science* 268 (April 28, 1995): 493.

65. Gabrynowicz, "Bringing Space Policy into the Information Age," 174.

66. Renee Marlin-Bennett, "Who Owns the Weather? A Case of Commodification in the Global Political Economy." Paper presented at the 1996 meeting of the International Studies Association, 17–21 April 1996, San Diego, 2. (cited with permission of author).

67. David Rhind, "Geographical Information Systems and Environmental Problems," *International Social Science Journal* 43 (November 1991): 662.

68. L.A. Fisk, L.K. Berman, R. Brescia, A.S. McGee, and F.C. Owens, "NASA Takes Lead Role in Earth Observation," *IEEE Technology and Society Magazine* (Spring 1992): 11.

69. Thomas W. Becker, "Mission to Planet Earth and Global Space Education Policy," *Space Policy* 8, no. 2 (May 1992): 158. Many ERS enthusiasts also advocate a crash program in the developing countries aimed at "creating a scientific outlook." See U.R. Rao, chairman of India's Space Commission, "Space Technology in Developing Nations: An Assessment," *Space Policy* 9, no. 2 (May 1993): 169.

70. Stewart, "Eyes in Orbit Keep Tabs," 71.

71. Graham Harris, "Global Remote Sensing Programmes, Global Science, Global Change: An Australian Perspective," *Space Policy* 9, no. 2 (May 1993): 131. A current example is the struggle that the new one-meter systems face in winning customers who currently use airplanes for mapping or monitoring. Aerial remote sensing involves "well-ensconced relationships with a local pilot, a local customer, [and] local analytical people." See Joseph Anselmo, "High-Resolution Satellite Competiton Heats Up," *Aviation Week and Space Technology* 140 (July 11, 1994): 56.

72. For instance, see Scientific American, *Managing Planet Earth* (New York: W.H. Freeman and Co., 1990).

73. James R. Beniger, "Information Society and Global Science," *Annals of the American Association of Political and Social Scientists* 495 (January 1988): 23.

74. Daniel Clery and William Bown, "Sensing Satellites: Who Calls the Tune?" *New Scientist* 130, 1767 (May 4, 1991): 17.

75. Hatoyama-Machi, "Japanese Earth Satellites," 71.

76. Lindgren, "Commercial Satellites Open Skies," 36.

77. Martua Sirait, et al., "Mapping Customary Land in East Kalimantan, Indonesia: A Tool for Forest Management," *Ambio* 23, no. 7 (November 1994): 411–17; Cultural Survival, *Geomatics: Who Needs It?* 18, no. 4 (Spring 1995); Jefferson Fox, ed., "Spatial Information and Ethnoecology: Case Studies from Indonesia, Nepal, and Thailand," an unpublished manuscript available from the East-West Center in Honolulu, Hawaii. I am grateful to Judith Mayer for her insight into the reterritorializing of political practice at the grass roots level.

78. Masahide Kato, "Nuclear Globalism: Traversing Rockets, Satellites, and Nuclear War via the Strategic Gaze," *Alternatives* 18 (1993): 344–45.

79. Beaumont Newhall, *Airborne Camera: The World from the Air and Outer Space* (New York: Hasting House, 1969): 121.

80. This is exactly the kind of issue that bedevils efforts to measure responsibility by formulating a "greenhouse index." See Peter Hayes and Kirk Smith, eds., *The Global Greenhouse Regime: Who Pays? Science, Economics, and North-South Politics in the Climate Change Convention* (Tokyo: Earthscan, 1993).

81. Jefferson Fox, "Spatial Information Technology and Human-Nature Relationships," East-West Center Working Paper No. 43, (Honolulu: East-West Center, 1995).

82. Susan Sontag, *On Photography* (New York: Farrar, Straus, and Giroux, 1977): 176–77.

83. P.M. Banks and C.C. Ison, "A New Role for Freedom," *Aerospace America* (September 1989): 30.

84. Carol Matlack, "Landsat's Slow Death," *National Journal* (July 25, 1987): 1903.

85. E. David Hinkley and Gary T. Rosiak, "Instrumenting Space Platforms for Earth Observations," in *Proceedings of the Pacific Rim Environmental Research Meeting* (World Scientific Publishing Co. PTE, 1992).

86. Fisk, "NASA Takes Lead Role," 14.

87. Christine Nielsen and Dirk Werle, "Do Long-Term Space Plans Meet the Needs of Mission to Planet Earth?," *Space Policy* 9, no. 1 (February 1993): 15.

88. World Meteorological Organization, *Global Climate Observation System: Responding to the Need for Climate Observations* (Geneva: WMO, 1992): 12.

89. Ruggie's notion of a new social episteme entails "a new set of spatial, metaphysical, and doctrinal constructs through which the visualization of collective existence on the planet is shaped." Ruggie, "Territoriality and Beyond," 173–74.

9

Sovereignty, Environment, and Subsidiarity in the European Union

Joseph Henri Jupille

Introduction

Taken separately, sovereignty and environment in the European Union (EU) have spawned innumerable scholarly works. From the 1950s onward, the "uniting of Europe" has put sovereignty issues high on the political and academic agendas, making European integration literature a hotbed of speculation on the fate of sovereignty in a wider context. More recently, environmental interdependence and the growth of an EU environmental policy have served to nourish another burgeoning literature concerned principally with environmental policy integration. However, despite a growing body of literature in international relations dealing with the mutual influences of sovereignty and environment, to date very few studies have explicitly examined this connection in what would seem to be the most natural of cases, the European Union.[1] The confluence of the continuing problematization of sovereignty and innovative responses to environmental needs makes the EU a useful case to examine in assessing the interplay between sovereignty and environment more generally.

In this chapter, I first briefly outline current usages of sovereignty, which, I argue, conflate numerous concepts that might be more analytically useful when considered in isolation. Second, I pick up on a theme first articulated by Joseph Weiler, which could be paraphrased here as the "dual character" of sovereignty in the European Union: Whereas legal analyses attest to the growth of supranationalism and the diminution of state sovereignty, political analyses underscore the continuing relevance of states.[2] Within this framework, then, I trace the contours of the EU's response to environmental needs, focusing on its implications for state sovereignty. In the concluding section I turn to the notion of *subsidiarity*, or the idea that policy will be enacted at the level most appropriate to the problem being addressed, which I portray as a response to environment that moves beyond sovereignty yet resonates with themes emanating from the sovereignty/environment discourse.[3] I conclude that the EU's environmental responses entail innovations that cause us to rethink the interrelationship between political authority and ecological needs.

Sovereignty as it has come to be used in modern scholarship can mean many things. At a maximum, it is "the absolute and perpetual power of a commonwealth,"[4] as Jean Bodin put it in 1577. In more contemporary parlance, this power-based conceptualization might simply mean "a nation's practical capacity to maximize its influence in the world."[5] F.H. Hinsley offers a similar, but authority-driven, conception in defining sovereignty as the "final and absolute authority in the political community."[6] Finally, a minimalist stance equates sovereignty with simple independent decision-making ability: In Waltz's formulation, "To say that a state is sovereign means that it decides for itself how it will cope with its internal and external problems."[7] All of these usages either explicitly or implicitly delimit sovereignty spatially. Be they in reference to the state or the more generic "political community," the rights, powers, or responsibilities described only pertain within a given territory. While I will not explicitly feature territoriality in the discussion of sovereignty and environment in the EU, its relative lack of importance with respect to subsidiarity will be highlighted in the final section.

It follows from the above that our current usages of *sovereignty* are imbued with multiple interdependent elements.[8] This commingling of distinct concepts hinders efforts at talking commensurably about the impacts of such phenomena as environment, human rights, technology, or economic interdependence upon sovereignty (or its impact upon them). In this chapter, I unpack the prevailing usages of sovereignty and attempt to disentangle three central components: authority, which I employ in the sense of the legal competence to take decisions, rather than the broader notion of authority adopted by others in this volume; power, or the ability to alter outcomes; and autonomy, or power plus independence.[9] These three concepts are highly intertwined, which is

Table 9.1
Sovereignty and environment: International and EU responses

	Authority	Power	Autonomy	Nonsovereign
International	No effect	Increase	Decrease	None
EU	Decrease	Increase	Decrease	Subsidiarity

perhaps why they have been "bundled" together into sovereignty in the first place. However, it is analytically useful to separate them because separation allows us to see that environment and sovereignty are not necessarily either mutually destructive or mutually supportive, but rather influence each other in multiple and varied ways.

Legal authority as used here does not have to be operational, but can exist "equally" even where power varies wildly. The juridical equality of states has never necessarily entailed equality of condition or power. Further, legal authority does not necessarily entail or even require autonomy. In situations of so-called pooled sovereignty, as many argue applies in the European Union, it may be the case that actors are legally required to take decisions in common.[10] They cannot, under such conditions, be said to be autonomous. Power can exist with or without legal authority and autonomy, whereas autonomy can exist without authority, but subsumes at least some notion of power plus independence. Depending on the situation, environment can affect all of the above in myriad ways.

By unpacking sovereignty, I intend first and foremost to clarify the mutual effects of environment and sovereignty in the EU. I argue that environmental response has employed, but also generated, changes in the EU that are different from those seen in the international system more generally (see Table 9.1). International responses to environmental exigencies reduce the autonomy of political actors, increase their power, and almost always leave (state) authority intact. International environmental responses take the form of joint choices that preclude full autonomy for individual states, although they increase state power to the extent that they provide leverage on problems that cannot be achieved unilaterally.[11] Authority remains intact internationally, however, as states usually retain the general right of exit as well as the right of veto over any particular decision. Internationally, states enter into and remain parties only to agreements of their choosing.[12]

In the EU, the effects of environmental responses on state power and autonomy work in the same direction as they do internationally, although they may have a different intensity because of the sheer depth of political and ecological interdependencies in Europe. The first difference between the EU and international responses to environment, however, is that Europe has in some ways employed and effected a modification of legal authority, so that under certain conditions, EU member states can be bound by environmental measures to the passage of which they did not consent.[13] The second innovation in the EU is its use of the notion of subsidiarity to guide its environmental response. Subsidiarity, or the idea that policies should be pursued at the level most appropriate to the problem being addressed, in many ways transcends the sovereignty-derived categories of authority, power, and autonomy, and severs the explicit relationship between territory and political authority so fundamental to sovereignty. Subsidiarity integrally links legal authority to function and democracy in a dynamic, rather than a codified and static, relationship. Of the two innovations evident in the EU, it is argued here that subsidiarity may have the more profound implications for sovereignty and environment. Although the modification of legal authority is a novel response and is arguably unique to the European Union, it is in many ways counteracted by continued power and autonomy for EU member states. Subsidiarity is a new type of response that relies neither on legal procedures nor on power and autonomy, acting instead as a general ordering principle that guides, but does not dictate, the dispersion of political authority.

Modifying Legal Authority

The legal authority of the EU and its member states in environmental matters has evolved over four distinct phases.[14] The entry into force of the 1997 Amsterdam Treaty would bring about a fifth stage in this development. The first phase, from 1958 to 1972, was one of a total absence of community competence (or even of a community policy per se) in the environmental sphere. The founding Treaty of Rome provided no explicit legal basis for environmental action by the European Economic Community (EEC), and the few environmentally relevant measures that were enacted were generally undertaken on an ad hoc basis to facilitate the completion of the Common Market. Late in this period, officials began reflecting on the need to establish an EU-level policy, but little thought was given to the establishment of an explicit community competence by any institution other than the European Parliament (EP).[15] Even after the leaders of the EU member states gave the political go-ahead to community-level environmental action at the 19–20 October 1972 Paris Summit,[16] there was no real concept of a member state loss — or explicit community gain — of legal authority to respond to environmental exigencies.

The second period, from 1972 through 1986, saw some movement in the development of a European competence in the environmental sphere, although again this entailed no modification of member states' legal authority. One insider has said of this period that environmental policy was "reactive, incidental, and unarticulated."[17] As there was no explicit legal basis for community action in environmental affairs, EU-level action required the member states to creatively read such competence into the Treaty of Rome. This was done with the double legal basis of Articles 100 and 235, the first of which dealt with the Common Market and called for approximation of national laws, and the latter of which was a residual "implied powers" clause for the community, to be used for environmental measures that would in no way pertain to the Common Market.[18] Both articles employed unanimity voting, so member states effectively retained the right of veto over any environmental measures foreseen for the community level.

The third period encompasses the time between the entry into force of the Single European Act (SEA) in 1987 and that of the Treaty on European Union (Maastricht Treaty) in 1993. The SEA, the raison d'être of which was the completion of the Single European Market through the "1992" program, granted community environmental action explicit legal status by adding to the Rome Treaty (Articles 130r-t) a "title" that was concerned solely with the environment.[19] The new articles foresaw the formulation by unanimity of common environmental policies (130s) and allowed for the maintenance or enactment of "more stringent protective measures" by the member states so long as they were compatible with the treaty's other provisions such as the free movement of goods (130t).[20] Article 130r(4) introduced an implicit "subsidiarity clause" which authorized community environmental action only where it was the most appropriate level of policy response. While giving the EEC the legal right to undertake environmental actions, virtually the entire environment chapter preserved and even bolstered member states' legal authority, not only by maintaining the unanimity requirement and thus the national veto, but also by allowing for more stringent national measures under Article 130t.

Crucially, though, the SEA also enacted changes that could allow for a modification of state legal authority in the pursuit of environmental objectives where these were explicitly linked to the completion of the Single European Market. The newly added Article 100a, which dealt with harmonization measures (including environmental ones) needed to complete the Single European Market, called for qualified majority voting, thus obviating the possibility of a single member veto and raising common minimum standards above the lowest common denominator.[21] Paragraph 4 of the same article, however, maintained unilateral authority for higher levels of environmental protection. At the insistence of ultragreen Denmark, this provision allowed member states that had been outvoted by the qualified majority to continue to apply existing "national provisions" (stricter standards) in order to protect the environment.[22] Here the EU solution is truly novel. In addition to maintaining the authority needed to protect high national standards (as implicitly allowed for in most international agreements), under Articles 130t and 100a(4), the Single European Act removed the national veto as an impediment to the attainment of many common minimum standards. The move to qualified majority voting is an important modification of the legal authority component of sovereignty and has led some observers to view the EU as a postmodern political form.[23]

The fourth period, the current one, has been heavily influenced by the Maastricht Treaty, in force since November 1993.[24] Although at least one analyst holds that Maastricht "does not herald radical changes in the environmental field," others have asserted that it "represents a significant strengthening of [EU] environmental policy," bringing about greater decision-making efficacy and further modifying state legal authority by extending the use of qualified majority voting in the Council of Ministers to most areas of environmental policy-making under the environment title.[25] For our purposes, however, it suffices to note that the trend toward the modification of legal authority begun under the Single European Act was extended under Maastricht, and that, henceforth, qualified majority voting applies to all environmental questions except those "primarily of a fiscal nature," those concerned with town and country planning, land use, or the management of water resources, and those which might alter the structure of national energy markets.[26]

In sum, the need to respond to environmental exigencies has led to a modest diminution of state sovereignty in the European Union to the extent that individual member states can be bound by measures to the passage of which they did not consent. As Geoffrey Palmer notes, such "legislative" capacity is fundamentally different from the treaty-making capacity characteristic of international environmental agreements.[27] Where states can enact unilateral measures despite common community standards, the measures are subject either to a compatibility test with other community rules, to approval by the European Commission, or both.[28] The move toward qualified majority voting and the concomitant abandonment of the national veto are critical indicators of the modification of legal authority, and the move toward what Weiler calls "normative supranationalism." However, as the unbundled concept of sovereignty used here suggests, power and autonomy must in their turn be considered if we want to develop a more nuanced and useful understanding of the sovereignty/environment nexus in the EU.

Power and Autonomy

Although one can differentiate the EU from "traditional" international environmental responses to the extent that it modifies legal authority, the direction of its influence on the other two unbundled components of sovereignty—power and autonomy—is the same as on the international scene more generally. Prevailing analyses of the nature of sovereignty in the EU, however, tend to privilege either legal phenomena or their political counterparts, and in so doing draw inaccurate or incomplete conclusions. Legal scholars tend to focus on the growth of the EU's competence in the environmental sphere and on the relative diminution of the states' authority. Given that "normative supranationalism" seems to be on the rise, these scholars argue that sovereignty is on the wane.

Political scholars, on the other hand, tend to focus on the continuing relevance of the member states in checking "decisional supranationalism," and draw the opposite, but similarly flawed, conclusions.[29] Yet both the legal and the political aspects are important to understanding sovereignty and environment in the EU.

Power: Policy Formulation and Decision Making

State power, or ability to alter policy outcomes, remains substantial in the European Union despite the modification of legal authority. This resiliency of state power is especially manifest in states' influence in environmental policy formulation and decision making, which I will describe briefly here before turning to a fuller examination of state autonomy.

As noted, internal EU environmental measures can be based principally on Article 130s and Article 100a, which foresee the use of different legislative procedures, each entailing different levels and forms of involvement for the member states and the EU institutions. Very generally, however, member states maintain considerable informal influence in the prelegislative stages of policy formation. Despite the European Commission's formal monopoly on the right of policy initiative, national representatives are almost always sounded out by the commission in order to determine their home governments' stances on prospective policies.[30] In addition, even though the treaty now explicitly calls for qualified majority voting in most areas of environmental policy-making, the Council of Ministers still frequently operates on a consensus basis, so that any conclusions we draw about the effects of modifications in legal authority on member state power should be strongly qualified.[31] Having asserted this, it is still useful to understand the extent of state involvement under formal decision-making procedures.

With the entry into force of the Maastricht Treaty, environmental measures undertaken on the basis of Article 130s are to be legislated using the so-called cooperation procedure found in Article 189c, under which the commission proposes legislation, the Council of Ministers reaches a "common position" by qualified majority vote, and the European Parliament can do one of three things. First, if the EP accepts the council position, the legislation is passed. Second, if it rejects the common position, this rejection can only be overridden by unanimous council vote. Third, if the EP amends the position, the proposal goes back to the commission, which can adopt or ignore any or all of the amendments. When the commission accepts EP amendments, the council may pass the legislation by qualified majority vote (QMV) or amend and pass it unanimously. When the commission rejects EP amendments, the council must pass the legislation unanimously or else it lapses. Under this procedure, member states can alter policy outcomes in negotiations with fellow member states, the EP, and the commission, although rarely can they do so unilaterally.

Where environmental legislation is adopted under Article 100a (internal market provisions), it is subject to the "codecision" procedure which is laid out in Article 189b of the EEC Treaty as amended by Maastricht. This procedure is incredibly complex and cannot be dealt with at any length here. In lieu of a description, therefore, I make two points: First, there is currently a heated debate regarding the impact of the codecision procedure on the relative power of the European Commission, the European Parliament, and the member states in the Council of Ministers. Garrett, Tsebelis, and others claim that the commission and parliament are weaker under codecision than under the cooperation procedure.[32] Scully, Jacobs, and others contend that codecision represents an advance in the powers of the parliament relative to the cooperation procedure and puts it on more equal footing with the Council of Ministers.[33] Clearly this remains an open debate.[34]

Second, whatever the result of this debate, it is useful to note that procedural changes themselves come about because of explicit and unanimous member state choices. As with the Single European Act and its innovative cooperation procedure, the Maastricht Treaty and the codecision procedure resulted from an intergovernmental conference of the member states and was unanimously signed and ratified. Member states expanded the scope of the codecision procedure in the 1997 Amsterdam Treaty. Therefore, the definition of procedural alternatives according to which issues are processed also becomes a potential source of state leverage, although the fact that this can only take place unanimously at a formal intergovernmental conference (IGC) makes it a high hurdle. Where legal authority has been modified, there is still considerable direct and indirect scope for state influence on environmental policy outcomes.

Autonomy: National Standards and Implementation

The third unbundled element of sovereignty, much like the second, also attests to the continuing relevance of EU member states. State autonomy, meant here as power (ability to alter outcomes) that can be wielded independently, remains considerable even in the highly integrated European Union. While numerous aspects of EU policy-making attest to continuing state autonomy, this dimension of the sovereignty/environment nexus is most prominent in two areas: member states' ability to maintain or enact national standards different from the EU norm, and the national implementation of EEC-level policies and the "postdecisional politics" entailed.[35] The first will be treated briefly before a lengthier examination of the second.

National Standards   Despite efforts at achieving common standards, member states have considerable leeway in determining their own levels of environmental protection. State autonomy of this sort has come about in two ways. First, member states have consciously poked holes in both the EU regulatory ceiling and the regulatory floor. As noted, the SEA allowed member states to maintain stricter environmental standards under Articles 100a and 130t. The latter also allowed members to enact new standards stricter than the community norm even after adoption of EU measures. With Maastricht, member states opened up holes in the regulatory floor by allowing low-standard states to derogate temporarily from higher common standards.[36]

Second, the European Court of Justice has gradually allowed member states greater scope to enact unilateral environmental measures despite the fact that they may restrict the free movement of goods. In its landmark Cassis de Dijon judgment, the court had seemingly vitiated state autonomy in the area of protection from lower standards by establishing what has come to be known as the doctrine of mutual recognition, according to which goods lawfully marketed in one EU member state cannot legally be prevented from entering other member states' markets. However, the court established a safeguard, now called the "rule of reason," according to which member states seeking to satisfy certain "mandatory requirements" such as the protection of public health or the defense of public morality could restrict free trade.[37] In its equally important 1988 Danish Bottles judgment, the court found that environmental protection constitutes one such "mandatory require-ment" and could be used by member states to justify high standards that had the ancillary effect of restricting free trade.[38] More recently, the court has gone so far as to allow member states to maintain environ-mental measures that not only restrict trade, but are on their face discriminatory.[39] In sum, this line of jurisprudence has broadened the scope of member state autonomy by affording states greater leeway in the unilateral enactment of environmental measures.[40]

Implementation   Through various legislative procedures, by 1995 the EU had adopted a total of 179 directives, 71 regulations, and 41 resolutions as the body of its environmental policy.[41] While the process by which these acts were formulated varies according to the legislative procedure in use (which is a function of the treaty article on which they are based), the modality of their implementation is given by the type of legislative instrument employed. Regulations are directly applicable in and immediately binding on member states — they are transposed verba-tim into national law — whereas the more prevalent directives are specific as to the goal to be achieved, but silent as to the means of achieving them. With directives, legislation is implemented at or below the state level. Because the legislative implementation of directives takes place in the national arena, both the executives and the legislatures of member states retain considerable unilateral ability to influence environ-mental policy outcomes.

Although implementation is often described as the Achilles' heel of EU environmental legislation,[42] in quantitative terms the record of member state transposition of EU directives seems largely satisfactory. Complete nonimplementation is relatively rare (see Table 9.2). In qualitative terms, however, the record is much more mixed, with inconsistent implemen-tation being perhaps "the biggest problem with [EU] environmental legislation."[43] Thus, in addition to the first form of implementation difficulties (nonimplementation), two others must be considered: the failure by member states to completely and correctly transpose commu-nity obligations into national law, and "practical nonimplementation" (i.e., failure to respect the spirit of community legislation), which is a more telling example of the continued and sometimes conscious main-tenance of state autonomy.[44]

Table 9.2
National transposition of EU environment directives, end 1995

Country	Directives applicable	Measures notified	Percent
Belgium	133	111	83
Denmark	133	131	98
Germany	135	127	94
Greece	138	123	90
Spain	137	111	86
France	133	126	95
Ireland	133	127	95
Italy	133	113	85
Luxembourg	133	122	92
Netherlands	133	131	98
Austria	129	119	92
Portugal	137	119	87
Finland	131	114	87
Sweden	131	123	94
United Kingdom	125	102	82

Source: Commission of the European Communities, COM(96) 600 final, Thirteenth Annual Report on Monitoring the Application of Community Law (1995) (29 May 1996): 91.

The potential for practical nonimplementation is endemic to the EEC system because it is particularly hard to observe, because the EU Commission has few resources with which to monitor the application of environmental law by member states, and because the community has traditionally had little recourse against nonimplementation even when it is observed.[45] However, in addition to the existing infringement proceeding under Article 169 of the Treaty of Rome, the Maastricht Treaty did grant some power in this respect to the commission and the European Court of Justice. With Maastricht, the court gained the ability under Article 171 to "impose a lump sum or penalty payment" on member states that do not comply with their obligations, on the recommendation of the commission.[46] As of mid–1996, Article 171 proceedings related to the environment were underway against eight of the EU's fifteen member states, and in 1997 the commission proposed significant penalties against both Germany and Italy for failing to remedy infringements of EU environmental law, compelling them to adjust their implementing legislation in line with commission wishes.[47]

In addition to underscoring the EU institutions' traditional weakness vis-à-vis the member states, practical nonimplementation also highlights the extent to which modifications of legal authority can incite surreptitious member state reactions. In the case of the extension of qualified majority voting by the Maastricht Treaty, for example, member states that oppose enacted legislation may be more tempted than ever simply not to implement community directives.[48] Because they are the only ones who can implement most community legislation, member states retain a considerable ability to manipulate environmental policy outcomes independently of the EU institutions and their fellow members.

It would be misleading, though, to assert that nonimplementation is solely the product of member state willfulness. The European Parliament's Environment Committee attributes the nonimplementation of EU environmental legislation to a number of structural factors.[49] First, the large volume, imprecision, rigidity, and poor integration of environmental legislation into other policies all make it an unwieldy body of policy to implement. Second, the functioning of the community institutions themselves plays a role. Decision making is cumbersome and inflexible, and community institutions such as the commission's Directorate General for Environment (DG XI) and the EP's Environment Committee are understaffed and underfunded.[50] They are also poorly coordinated, with numerous Directorates General — e.g., for agriculture or industry — responsible for various aspects of the EU's environmental policy.[51] Third, economic, cultural, and sociological factors in the member states condition national acceptance of community environmental policy. Low environmental standards favoring the cost-competitive positions of domestic producers, high standards which serve as barriers to the entry of foreign products, ineffective use of fiscal and tax incentives, and differing levels of ecological awareness or concern all shape member states' willingness and ability to implement their EU obligations. Finally, the administrative capacities of the member states can impede the implementation of EU environmental legislation, especially where the states themselves are nonunitary.

The strategy envisaged by the EEC and others to strengthen compliance with EU environmental law, and thus to reduce national autonomy in environmental responses, relies on actors below, across, and above the states as well as on the states themselves. It thus attests to the growing dispersion of legal authority, recognizing both the continuing relevance of states and, concurrently, the presence and relevance of other levels of political activity.[52]

Below and across the state—generally speaking, closer to the citizens—the European Union Commission has provided financial and administrative support and information to numerous NGOs, including the European Environment Bureau (EEB) and Friends of the Earth, to enable them not only to inform citizens about EU environmental law, but also to monitor its application themselves.[53] The European Union Commission has also sought to reinforce the environmental role of regional authorities by establishing a permanent regional information system which will allow them input into the environmental policy-making process, and it has provided them with financial support so that they can aid the commission in monitoring the implementation of environmental legislation.[54] The European Court of Justice has greatly expanded individuals' ability to challenge legally their states' failure to properly implement legislation through its extensive case law on the "direct effect" of EU rules.[55]

Recognizing that citizen complaints and European Parliament questions have been the most influential forces in aiding it to detect improper application of EU environmental legislation, the EEC has also sought to provide individuals and groups with better access to environmental information with Directive 90/313/EEC, which requires that reasonable requests for information be honored in a timely fashion.[56] To the extent that the commission is "almost entirely dependent" on citizens to bring allegations of nonimplementation to its attention, it has attempted to provide them with better financial and informational means to serve as its eyes and ears.[57]

At the level of the state, Directive 91/692/EEC on the harmonization of national reports gives the community access to cross-nationally comparable data, allowing it to judge states not only on the basis of citizen and EP complaints, but also on states' own information.[58] The member states and the community also participate in an informal implementation network, set up under the Fifth Environmental Action Program, under which working groups meet to discuss specific problems of implementation. Plenaries representing the commission and the member states meet to discuss these working group reports.[59] The community intends to bring together in this transnational forum all of those involved in implementation, from policymakers to regulators and administrators, so that they can share experiences and expertise with an eye toward systematizing implementation.

Finally, above the state, the EEC has created the European Environment Agency (EEA) and the European Environment Information and Observation Network, which are charged with gathering objective, reliable, and comparable information in order to facilitate implementation and public awareness.[60] Some actors had wanted, and continue to push for, more invasive monitoring and investigative capabilities for the agency, but for the time being its role is limited to that of data collection and analysis.[61] A middle road in this regard might be adoption of the suggestion put forth by the British House of Lords to create at the EEC level an "Inspectorate of Inspectorates," not itself responsible for monitoring implementation, but for overseeing the monitoring activities of national and subnational authorities.[62]

There has also been much discussion among both unofficial and official observers of strengthening the European Court of Justice's sanctioning capability. Instead of only being able to impose a lump-sum or penalty payment on an offending member—and then only on the recommendation of the commission, as is currently the case—the court might be able to impose fines itself and would be empowered to order the suspension or cancellation of EU funds otherwise due to the state in question.[63] All of the above measures would serve to lessen member states' ability to independently alter environmental policy outcomes, would thus lessen their autonomy, and as such would give further notice of the erosion of sovereignty in response to environmental needs.

In sum, the preceding two sections demonstrate that the sovereignty/environment question in the EU is not an either/or proposition. The member states' legal authority has indeed been eroded, both independent of environment and, more important for this paper, in direct response to it with the extension of qualified majority voting to most areas of environmental policy-making. However, the continuing relevance of the two other components of sovereignty which I have identified—power and autonomy—compels a more nuanced understanding of the sovereignty/environment nexus. Member states retain considerable influence on environmental policy outcomes through their insinuation in EU policy formulation and decision making, and they retain considerable autonomy through national standards and their preponderance at the implementation stage. What emerges is a complex picture of sovereignty and environment in the EU, one that requires continued attention to both the European and national levels, as well as to the legislative procedures and instruments employed in response to perceived environmental needs. Despite these innovations, I contend that one further development in the EU warrants even more scrutiny on the part of those wishing to understand the interplay between environment and political order: the rising importance of subsidiarity as a guide to environmental responses and a potential alternative to sovereignty as a way of conceptualizing political authority.

Beyond Authority, Power, and Autonomy: Subsidiarity

Subsidiarity, or broadly speaking the notion that policies should be pursued at the level most appropriate to the problem being addressed, presents a way of imagining political responses to environmental needs which is not encumbered by any explicit relationship to the notion of sovereignty, either as that term has come to be used or in the unbundled way in which it is employed here.[64] Although it has been employed in the EU as a way of conceptualizing a structure that is neither a federal system nor a simple international organization, and thus has no necessary connection to the environment, it has come to be used perhaps most often in connection with the EU's environmental policy, as a guide to which level of government should address which environmental problems in the absence of a constitution specifying the division of competencies.[65]

Subsidiarity, with strands of classical, Calvinist, and Enlightenment thinking, is not a new idea, but it apparently first was used in this century by Pope Pius XI in his 1931 Encyclical *Quadragesimo Anno*.[66] In the European Economic Community context, the idea emerged in the European Parliament in the 1970s as both a reflection of the dominance of Christian democratic thought in that body and as a potential solution to the perceived inability of states to respond to social, environmental, educational, industrial, and macroeconomic challenges. At that time, then, it was seen as a tool by which EU competencies could be expanded, using efficacy as a criterion, in order to effect policy goals which were seemingly beyond the capabilities of the constituent member states.[67]

As it emerged from the 1970s, however, subsidiarity began to take on more overtly decentralizing overtones. In the European Parliament's ill-fated 1984 Draft Treaty on European Union, subsidiarity was presented by the treaty's architect (and ardent federalist) Altiero Spinelli as an explicit defense against the overcentralization of the community,[68] although the treaty did undoubtedly set up an efficiency/feasibility criterion for the allocation of authority rather than a constitutional/political one. Under Article 12 of the treaty, which was never enacted (although it did affect later developments including the Maastricht deliberations on subsidiarity), the "union" would "only act to carry out those tasks which may be undertaken more effectively in common than by the Member States acting separately, in particular those whose execution requires action by the Union because their dimension or effects extend beyond national frontiers."[69] When read in conjunction with Article 59, which would empower the Union to enact environmental measures to prevent or remedy damage which was "beyond the capabilities of the individual Member State or which requires a collective solution,"[70] the implication of the Draft Treaty is that sovereignty would no longer be used to determine the "level" of environmental responses in the Union.

The first binding, albeit implicit, usage of *subsidiarity* came in the Single European Act's Article 130r(4), in which it was applied only to environmental action. This article limited community environmental action to cases in which the EU's environmental objectives could "be attained better at Community level than at the level of the individual Member States."[71] In the Maastricht Treaty, subsidiarity was moved from the environment chapter to the "principles" section, and was thus generalized to apply to the entire range of EU policies. Under Maastricht's Article 3b, the EU is allowed to take action "only if and insofar as the objectives of the proposed action cannot be sufficiently achieved by the Member States and can therefore, by reason of the scale or effects of the proposed action, be better achieved by the community."[72] Subsidiarity is thus a general principle, one which seeks to limit central (EU-level) intervention to those policies for which it is necessary. Given this general conceptualization of subsidiarity, what light does it shed on the sovereignty/environment discourse, and how may it come to be used in guiding environmental responses in the EU context?

First, it is useful to examine the conceptual limitations of subsidiarity. There is widespread sentiment among academic observers and policy participants in Europe that subsidiarity is excessively malleable, leaving it susceptible to problems of both politicization and operationalization. The politicization of subsidiarity predates—indeed prefigures—its inclusion in the Maastricht Treaty. It came to first be employed by members of the European Commission (especially its President, Jacques Delors) from the mid–1980s onward to soothe popular fears of an overactive and unrepresentative Brussels bureaucracy arrogating to itself powers that might more properly remain with EU member states or subnational authorities. This politicization intensified in the wake of the Maastricht Treaty and the events of 1992—especially the Danish rejection of Maastricht by referendum in May of that year and the French *petit oui* in September—which called the viability of the Maastricht agreement, and indeed the entire EU structure, into question and brought subsidiarity to the fore.[73]

Article 3b of the Maastricht Treaty was originally included as a political and rhetorical compromise between those who sought to establish the EU's "federal vocation" and those, such as the British, who sought to undermine it. This demonstrates the extent to which subsidiarity is an eminently manipulable concept, capable of being all things to all people.[74] Peterson speaks of three ideologies of *subsidiarity*, each of which sees in the term a way of promoting its own vision of the EU. The first is the Christian democratic version, which stresses the virtues of small social groupings and a dynamic view of center-periphery relations; the second is the German federalist version, which stresses a static constitutional division of powers; and the third is the British conservative version, which stresses the autonomy of territorial (read: *central national*) governments.[75] Each of these three competing ideologies has a certain weight in the EU and has shaped efforts to put subsidiarity into practice.

The problems with operationalizing *subsidiarity* have run up against the inherent ambiguity of the term. In attempting to apply the principle of subsidiarity to the preparation, management, and control of EEC policies, the European Union Commission, in a reflection document which it was charged with drafting in the wake of the Danish referendum, admitted that in practical terms *subsidiarity* implies "the application of the simple principle of good sense" on community institutions.[76] In this same document, the EU Commission also listed the criteria which proposed legislation would have to satisfy in order to respect the subsidiarity principle. In line with subsidiarity's more federalist connotations, the commission foresaw the adoption of a codified "hierarchy of legislative norms" under which different types of community legislative acts would be classified and ordered. Such criteria and such a hierarchy would require the commission to review existing and pending legislation and legislative proposals for their conformity with the subsidiarity principle. Although the council expressed hostility to the notion of a hierarchy of norms at the 1992 Edinburgh European Council, the European Parliament continues to endorse the idea,[77] and the operationalization of subsidiarity was a major topic of discussion at the 1996–97 intergovernmental conference which led to the Amsterdam Treaty.

On the basis of these early efforts, the commission came up with a list of eleven existing and proposed pieces of legislation to be withdrawn, none of which pertained, despite fears of repatriation, to environmental policy.[78] As it now stands, the EU's environmental policy seems to be relatively safe from any subsidiarity-driven onslaught. Action at the EU level appears to be appropriate, given the transfrontier nature of many environmental problems, its utility in avoiding distortions to trade which would result from differing national environmental regulations, and its popularity with both citizens and the EU's most powerful member states.[79] As it is being operationalized, subsidiarity will not be used to call into question the body of existing EU law (the *acquis communautaire*), but it will serve as a "guiding principle" against which all legislation proposed by the commission, or amendments suggested thereto by others, will have to be explicitly justified. The commission will also have to submit annual reports on compliance with the principle to the European Parliament and the European Council.[80] Very generally, then, as it is being put into practice, *subsidiarity* is becoming something of a cognate for "better lawmaking," or "doing less, but doing it better."[81]

Other criticisms of subsidiarity can also be anticipated: for example, that it may be interpreted as at best idealistic and at worst a bad guide for environmental policy-making. Subsidiarity is arguably idealistic in the sense that it assumes that the use of reason can point clearly to which level of authority is most appropriate in response to any given problem. However, one of the defining characteristics of environmental challenges is the scientific uncertainty they engender, which both contributes to and is exacerbated by different culturally, economically, and socially promulgated valuations of environmental soundness. Furthermore, even were it possible to use "reason" to assess the most appropriate level of response, subsidiarity paints a picture that is apolitical. Who decides what the appropriate level is, and how is this decision enforced? As to the second criticism, subsidiarity may be no better than sovereignty in ensuring effective environmental policies. If sovereignty errs by defaulting to the sovereign state, subsidiarity may err by deciding on one level, as if all of the units at that level were equally capable of effecting change. Are countries, subnational regions, or localities necessarily functionally equal in their ability to address environmental problems? Even though subsidiarity could theoretically allow for national and, elsewhere, subnational responses to the same problem, this may in fact so complicate matters that one would be left wishing for the simplicity of territorially bounded sovereign (state) authority.[82]

Despite these difficulties, subsidiarity is still conceptually useful, and can be characterized as outside of the sovereignty paradigm for two reasons. First, subsidiarity fundamentally delinks politics from territory. While subsidiarity would seem to relate to sovereignty to the extent that it refers generally to authority, power, and autonomy, it is distinct from sovereignty in that it does so nonterritorially. According to the subsidiarity principle, authority, power, and autonomy are elements that may reside in as restricted a space as one individual and in as wide a space as the entire globe. Political authorities overlap and inhabit the same space, and the relationship between them is nonhierarchical—it is one of "interconnectedness rather than nestedness."[83]

Second, "what is at issue" with subsidiarity "is effectiveness."[84] Subsidiarity is permeated by the notion of efficacy, in a way that differentiates it from the fundamentally rights-driven notion of sovereignty. Subsidiarity is widely regarded as a way of improving on the EU's twin deficits relating to policy (deficits of capacity and implementation) and democracy.[85] Policy efficacy demands that the location of authority be largely problem-driven. Rather than pursuing policy at a given level and then discovering that it lacks administrative or technical capacity to handle the problem, the policy level would instead be chosen on precisely the basis of its satisfactory policy capacity. This might lead to a green formula for subsidiarity according to which "decisions would be made at the level which would ensure maximum environmental protection."[86] Generally speaking, one might expect centralization where transfrontier physical, economic, or "psychic" spillovers are present or where economies of scale may be achieved, and devolution to occur where environmental or regulatory costs are internal to a given locality.[87] Thus, while subsidiarity may call for an EEC approach to transfrontier waste shipments or to environmental issues of global importance, it might on the other hand rule out community action on such issues as the quality of (local) bathing water.[88] The issue would decide not just the level, but the very nature of the arena in which it would be addressed.

Subsidiarity's emphasis on "democratic efficacy" has to do with making governance more effective for its citizens, and according to one observer it is an explicit criterion for democratic governance.[89] The European Parliament's Environment Committee, in considering subsidiarity's relationship with environmental needs, came down early and often in favor of democratic and accountable decision making as a guide to the allocation of competencies, and defined *subsidiarity* as "the principle by which democratic involvement is maximized in policy-making, implementation, and enforcement, and by which decisions are taken at the most local level."[90] It is an oversimplification, however, to simply equate subsidiarity with devolution. This is frequently done and is seen as leading to a repatriation of environmental responsibility, by definition at the expense of the EU, and allegedly at the expense of the environment.[91] While subsidiarity does have an initial devolutionary impulse, and as such establishes something of a general principle by which order may be established in a multilevel political setting, this a priori devolutionary prejudice does not necessitate devolutionary conclusions; rather, either devolution or centralization may take place, depending upon the particular issue under discussion.[92] The focus on policy efficacy makes democracy under subsidiarity truly operational. "Control by the people" no longer begins and ends with those problems that can be addressed by the sovereign state. The dual focus on efficacy differentiates subsidiarity from sovereignty, which is blind either to the effectiveness or to the (democratic) legitimacy of political action. Instead of looking first to the source of legal authority, and giving it policy-making priority—as under the sovereignty paradigm—subsidiarity looks first to the arena in which policy can best and most legitimately be enacted and uses this to determine where legal authority should therefore lie. This, it is argued, may be the most critical innovation in the EU's repertoire of environmental responses.

This flexibility may make subsidiarity a useful concept when one is rethinking sovereignty in light of environmental needs. First, it avoids the reification that ensues from the traditional sovereignty discourse. As the EU Commission has noted, subsidiarity's "function is not to distribute powers," but rather it is to serve as a principle or a "rule of reason" by which to consider the nature of relations between lower and higher levels of political authority.[93] Second, it is a democratic form of governance, which in addition to its other benefits may be preferable to other forms from an environmental point of view.[94] Its twin tests of "comparative efficiency" and "intensity" seek both to ascertain the necessity of EU-level action and to limit this action to what is minimally required for the attainment of a given goal.[95] It thus distances decisions from citizens on an "only as much as needed" basis. Third, subsidiarity's legitimation of all political levels attempts to provide a framework for dealing with the twin pressures of globalization and localization. It seeks to provide for both the extranational "control mechanisms" and the subnational psychological functions that are necessary in the face of environmental exigencies but which are ill served by the institution of sovereignty.[96] As such, subsidiarity provides "a mechanism to encourage cooperative action between different levels of government, rather than to fix boundaries or delineate any rigid division of competencies."[97]

Conclusion

The European Union is a useful case for understanding the sovereignty/ environment nexus at once because the interplay between them is emblematic of their interrelationship in the global political system, at least among the Western industrialized states, and because the solutions the EU applies to this interplay are so innovative. I have made three fundamental claims in this chapter. First, the term *sovereignty* as it has come to be used is actually a jumble of related, but unspecified notions that can usefully be decomposed into three distinct concepts: legal authority, power, and autonomy. Second, the implications for sovereignty of the EU's response to environmental needs resonate with Weiler's distinction between "normative" and "decisional supranationalism." While the legal component of state sovereignty is indeed being eroded, partly in response to environmental exigencies, sovereignty's political components, power and autonomy, display a greater resiliency which precludes any either/or conclusions about sovereignty and environment in the EU. Third, the emerging concept of *subsidiarity*—the idea that policy should be enacted at the most appropriate level—provides an intriguing alternative to sovereignty in imagining political responses to environmental needs, indeed in imagining political order in and of itself. Subsidiarity is a territorially nonexclusive ordering principle which both recognizes and embraces the proliferation of political authority above and below the state, and which uses policy and democratic efficacy, rather than "rights," to discern the appropriate level of political activity.

I do not make the claim that subsidiarity is poised to replace sovereignty as the ordering principle of life in the international system or even in the EU. Sovereignty is not going to disappear, if only because it comprises a fundamental component of larger cultural myths and psychological self-definitions which scholars and policymakers are unlikely to shed in the near future.[98] Nor do I make the claim that within its realm of applicability subsidiarity is currently operational or necessarily immune to politicization. Rather, I present subsidiarity as an imaginable alternative to sovereignty, a possible avenue for reconciling humanity with its environment in a way that is at once authoritative, effective, and democratic. At worst, it may indeed fall victim to the EU's member states' attachment to the prerogatives of sovereignty, or to some other obstacle in the EU structure. At best, however, it may prove useful in Europe and exemplary for others, perhaps reconciling political pluralism with ecological holism. And it may indeed confirm the EU's status as a "social laboratory for the global future."[99]

Notes

1. See, however, Jonathan Golub, "Recasting EU Environmental Policy: Subsidiarity and National Sovereignty," in Ute Collier, Jonathan Golub, and Alexander Kreher, eds., *Subsidiarity and Shared Responsibility: New Challenges for EU Environmental Policy* (Baden-Baden: Nomos Verlagsgesellschaft, 1997): 35–56.

2. Joseph Weiler, "The Community System: The Dual Character of Supranationalism," *Yearbook of European Law* 1 (1981) (Oxford: Clarendon Press): 267–306.

3. See more generally the outstanding discussion of some of these issues in Paul D. Marquardt, "Subsidiarity and Sovereignty in the European Union," *Fordham International Law Journal* 18 (December 1994): 616–40.

4. Julian H. Franklin, ed. and trans., *Jean Bodin on Sovereignty: Four Chapters from the Six Books of the Commonwealth* (New York: Cambridge University Press, 1992): 1.

5. Geoffrey Howe, "Sovereignty and Interdependence: Britain's Place in the World," *International Affairs* 66 (October 1990): 678; see also Howe, "Sovereignty, Democracy, and Human Rights," *Political Quarterly* 66 (July–September 1995): 128–30.

6. F.H. Hinsley, *Sovereignty* 2nd ed. (Cambridge: Cambridge University Press, 1986): 25–26.

7. Kenneth N. Waltz, *Theory of International Politics* (Reading, Mass.: Addison-Wesley, 1979): 96.

8. For this point see Ken Conca, "Rethinking the Ecology-Sovereignty Debate," *Millennium* 23 (Autumn 1994): 705–8.

9. This approach is similar to that suggested by Arthur P.J. Mol and J. Duncan Liefferink, "European Environmental Policy and Global Interdependence: A Review of Theoretical Approaches," in J.D. Liefferink, P.D. Lowe, and A.P.J. Mol, eds., *European Integration and Environmental Policy* (New York: Belhaven Press, 1993): 26–28.

10. On the pooling of sovereignty see Robert O. Keohane and Stanley Hoffmann, "Institutional Change in Europe in the 1980s," in Keohane and Hoffman, eds., *The New European Community: Decision Making and Institutional Change* (Boulder, Colo.: Westview Press, 1991): 1–39; Neill Nugent, *The Government and Politics of the European Union*, 3d ed.

(Durham, N.C.: Duke University Press, 1995): 433–35; and Emil Joseph Kirchner, *Decision Making in the European Community: The Council Presidency and European Integration* (New York: Manchester University Press, 1992): 10–14.

11. This latter argument is explored in Karen Litfin, "Ecoregimes: Playing Tug of War with the Nation-State," in Ronnie D. Lipschutz and Ken Conca, eds., *The State and Social Power in Global Environmental Politics* (New York: Columbia University Press, 1993): 94–118.

12. See, however, the discussion of the "new practices of sovereignty" entailed by international environmental agreements in Peter M. Haas with Jan Sundgren, "Evolving International Environmental Law: Changing Practices of National Sovereignty," in Nazli Choucri, ed., *Global Accord: Environmental Challenges and International Responses* (Cambridge, Mass.: MIT Press, 1993): 401–29.

13. Wallace has argued that this is why the EU has occasioned such hostility on the part of the British even though the substantive effects of international interdependence (which leave "formal sovereignty" intact but lessen autonomy) may be as considerable as the EU's. See William Wallace, "What Price Independence? Sovereignty and Interdependence in British Politics," *International Affairs* 62 (Summer 1986): 367–89.

14. A broadly similar account is given in Jan Jans, "The Development of EC Environmental Law," in Gerd Winter, ed., *European Environmental Law: A Comparative Perspective* (Aldershot: Dartmouth, 1996): 271–76.

15. SEC(71) 2616 final, First Communication of the Commission about the Community's Policy on the Environment, 22 July 1971; EP Doc. 15/72, *Rapport fait au nom de la commission juridique sur les possibilités qu'offrent les traités communautaires en matière de lutte contre la pollution du milieu et les modifications qu'il faut éventuellement proposer d'y apporter*, 17 April 1972.

16. See the Final Declaration from the Summit in Commission of the European Communities, *Sixth General Report of the Activities of the Communities, 1972* (Luxembourg: Office for Official Publications of the European Communities, February 1973): 8, 12.

17. Laurens Jan Brinkhorst, "Keynote Address—The Road to Maastricht," *Ecology Law Quarterly* 20 (1993): 9.

18. John A. Usher, "The Gradual Widening of European Community Policy on the Basis of Articles 100 and 235 of the EEC Treaty," in Jurgen Schwarze and Henry G. Schermers, eds., *Structure and Dimensions of European Community Policy* (Baden-Baden: Nomos Verlagsgesellschaft, 1988): 25–36; Brinkhorst, "Keynote Address," 10–11.

19. On the SEA and the environment see Ludwig Krämer, *EEC Treaty and Environmental Protection* (London: Sweet and Maxwell, 1990); Dirk Vandermeersch, "The Single European Act and the Environmental Policy of the European Economic Community," *European Law Review* 12 (1987): 407–29.

20. "Single European Act," *Bulletin of the European Communities*, Supplement 2/86.

21. The extralegal exercise of the national veto remains a distinct possibility, despite the fact that the Luxembourg Compromise that informally codified it is allegedly moribund. See Anthony L. Teasdale, "The Life and Death of the Luxembourg Compromise," *Journal of Common Market Studies* 31 (December 1993): 567–79; and "Veto-Mania," *The Economist* (18 February 1995): 48–49.

22. "Denmark Wins Assurances from EEC Partners That National Laws Will Not Be Compromised," *International Environment Reporter* (12 February 1986): 29; Claus Gulmann, "The Single European Act — Some Remarks from a Danish Perspective," *Common Market Law Review* 24 (Spring 1987): 34–40.

23. Asbjorn Sonne Norgaard, "Institutions and Postmodernity in IR: The 'New' EC," *Cooperation and Conflict* 29 (September 1994): 268. See also John Gerard Ruggie, "Territoriality and Beyond: Problematizing Modernity in International Relations," *International Organization* 47 (Winter 1993): 139–74; James A. Caporaso, "The European Union and Forms of State: Westphalian, Regulatory, or Postmodern?" *Journal of Common Market Studies* 34 (March 1996): 44–48.

24. *Treaty on European Union* (Luxembourg: Office for Official Publications of the European Communities, 1992).

25. The first quote is from Rudiger Wurzel, "Environmental Policy," in Juliet Lodge, ed., *The European Community and the Challenge of the Future*, 2d ed. (London: Pinter Publishers, 1993): 188; the second is from Barbara Verhoeve, Graham Bennet, and David Wilkinson, *Maastricht and the Environment* (London: Institute for European Environmental Policy, August 1992): 33; see generally Ludwig Krämer, *E.C. Treaty and Environmental Law*, 2d ed. (London: Sweet and Maxwell, 1995).

26. *Treaty on European Union*, Article 130s(2): 59. It should, however, be noted that Maastricht also established the right of low-standard states to temporarily derogate from community standards where the latter would be "disproportionately" costly for them in Article 130s(5). Citing this article, the four areas of exclusion, and the right to maintain higher standards, Kelemen asserts that with Maastricht the member states actually do much to reinforce or safeguard their legal authority. See R. Daniel Kelemen, "Environmental Policy in the European Union: The Struggle between Court, Commission, and Council," in Brigitte Unger and Frans Van Waarden, eds., *Convergence or Diversity? Internationalization and Economic Policy Response* (Aldershot: Averbury, 1995): 319–23.

27. Geoffrey Palmer, "New Ways to Make International Environmental Law," *American Journal of International Law* 86 (April 1992): 259–83.

28. The European Court of Justice has begun to define the limits of Article 100a(4), allowing unilateral measures in the face of common standards,

since the entry into force of Maastricht. See Case C-41/93, *French Republic v. Commission of the European Communities*, [1994] *European Court Reports*, 1829ff.

29. Contrast Daniel Vignes, "L'aménuisement de la souveraineté des États-membres des Communautés Européennes et l'intégration régionale européenne," *Studia Diplomatica* 47 (1994): 49–78, with Michael G. Huelshoff and Thomas Pfeiffer, "Environmental Policy in the EC: Neofunctionalist Sovereignty Transfer or Neorealist Gate-Keeping?," *International Journal* 47 (Winter 1991–92): 136–58, and Michael Mann, "Nation-States in Europe and Other Continents: Diversifying, Developing, Not Dying," *Daedalus* 122 (Summer 1993): 115–40.

30. Ludwig Krämer, *Focus on European Environmental Law* (London: Sweet and Maxwell, 1992): 139.

31. Ken Collins and David Earnshaw, "The Implementation and Enforcement of European Community Environmental Legislation," in David Judge, ed., *A Green Dimension for the European Community: Political Issues and Processes* (London: Frank Cass, 1993): 225.

32. See George Tsebelis, "The Power of the European Parliament as a Conditional Agenda-Setter," *American Political Science Review* 88 (March 1994): 128–42; Geoffrey Garrett, "From the Luxembourg Compromise to Codecision: Decision Making in the European Union," *Electoral Studies* 14 (September 1995): 289–308; Garrett and Tsebelis, "An Institutional Critique of Intergovernmentalism," *International Organization* 50 (Spring 1996): 237–68.

33. Roger M. Scully, "Institutional Change in the European Union: Maastricht and the European Parliament," and Francis Jacobs, "Legislative Codecision: A Real Step Forward?" Papers presented at the Fifth Biennial Conference of the European Community Studies Association, Seattle, Wash., 29 May–1 June 1997.

34. For empirical examinations see David Earnshaw and David Judge, "Early Days: The European Parliament, Codecision, and the European Union Legislative Process Post-Maastricht," *Journal of European Public Policy* 2 (December 1995): 624–49; and Jacobs, "Legislative Codecision."

35. This term is taken from the early political analysis of implementation found in Donald J. Puchala, "Domestic Politics and Regional Harmonization in the European Communities," *World Politics* 27 (July 1975): 496–520.

36. Kelemen, "Environmental Policy in the European Union," 316–20.

37. See especially point 8 of the judgment in Case 120/78, *Rewe-Zentral AG v. Bundesmonopolverwaltung fur Branntwein*, [1979] *European Court Reports*, 662.

38. Point 9 of the judgment in Case 302/86, *Commission of the European Communities v. Kingdom of Denmark*, [1988] *European Court Reports*, 4630.

39. Case C-2/90, *Commission of the European Communities v. Kingdom of Belgium*, [1993] *European Court Reports*, 4431–81.

40. Tamara L. Joseph, "Preaching Heresy: Permitting Member States to Enforce Stricter Environmental Laws than the European Community," *Yale Journal of International Law* 20 (Summer 1995): 227–71.

41. European Parliament, Directorate General for Research, *Directory of the Most Important Community Legislative Measures in Environment Policy*, Environment, Public Health, and Consumer Protection Working Document Series no. W-13 (April 1995): 3.

42. See for example EP Doc. A3–0143/92, Report of the Committee on Legal Affairs and Citizens' Rights on the Eighth Annual Report to the European Parliament on Commission Monitoring of the Application of Community Law (Rapporteur: Bandrés Molet, 27 March 1992): 11; see also *Debates of the European Parliament* [hereafter *EP Debates*] no. 3–417 (6 April 1992): 25–32.

43. Jody Meier Reitzes, "The Inconsistent Implementation of the Environmental Laws of the European Community," *Environment Law Reporter* 22 (August 1992): 10524.

44. Richard Macrory, "The Enforcement of Community Environmental Laws: Some Critical Issues," *Common Market Law Review* 29 (1992): 352–55.

45. There are, however, a number of monitoring techniques available to the European Commission in addition to recourse to the European Court of Justice. See Ludwig Krämer, "The Implementation of Environmental Laws by the European Economic Communities," *German Yearbook of International Law* 34 (1991): 12–14; see also Collins and Earnshaw, "Implementation and Enforcement," 227–34.

46. *Treaty on European Union*, Article 171, 69.

47. "Environmental Law: 22 Court Rulings Still Not Enacted by Member States," *European Report* no. 2144 (29 June 1996); "The Commission for the First Time Seeks Financial Penalties against Noncompliance with Court Judgments," press release IP/97/63, 29 January 1997; "Commission Decides to End Legal Action against Germany and Italy for Nonrespect of Environmental Directives," press release IP/97/568, 26 June 1997.

48. David Wilkinson, "Maastricht and the Environment: The Implications for the EC's Environment Policy of the Treaty on European Union," *Journal of Environmental Law* 4 (1992): 233.

49. EP Doc. A3–0001/92, Report of the Committee on the Environment, Public Health, and Consumer Protection on the Implementation of European Community Environmental Legislation (Rapporteur: Vernier, 6 January 1992).

50. At the end of 1992, just under 50 percent of those working for DG XI were outside contractors. See Answer to Question no. H-0555/93 by Mr. Muntingh, *EP Debates* no. 3–431 (26 May 1993): 234.

51. Court of Auditors, Special Report no. 3/92 concerning the environment, together with the commission's replies, OJ C 245 (23 September 1992): 6–7.

52. Johan From and Per Stava, "Implementation of Community Law: The Last Stronghold of National Control?," in Svein S. Andersen and Kjell A. Eliassen, eds., *Making Policy in Europe: The Europeification of National Policy Making* (London: Sage Publications, 1993): 55–67.

53. DG XI alone was the single largest source of external funding for the EEB in 1994. See European Environment Bureau, *EEB Annual Report of Activities 1994* (Brussels: EEB, April 1995): 35. See also Dennis R. Palmieri, "Extending the Reach of the EU Commission: Eurogroup Subsidization and the Implementation Process," manuscript, University of Washington, December 1996. For a perspective that laments the weakness of environmental NGOs in the EU see Ludwig Krämer, "Participation of Environmental Organisations in the Activities of the EEC," in Martin Führ and Gerhard Roller, eds., *Participation and Litigation Rights of Environmental Associations in Europe: Current Legal Situation and Practical Experience* (Frankfurt am Main: Peter Lang, 1991): 129–39.

54. *Agence Europe*, no. 6616 (30 November 1995): 11.

55. P.P. Craig, "Once upon a Time in the West: Direct Effect and the Federalization of EEC Law," *Oxford Journal of Legal Studies* 12, no. 4 (1992): 453–79.

56. Council Directive of 7 June 1990 on the freedom of access to information on the environment (90/313/EEC), OJ L 158 (23 June 1990): 56.

57. Collins and Earnshaw, "Implementation and Enforcement," 231.

58. Council Directive of 23 December 1991 standardizing and rationalizing reports on the implementation of certain directives relating to the environment (91/692/EEC), OJ L 377 (31 December 1991): 48–54.

59. Written question no. E–80/95 by Carmen Diez de Rivera Icaza, OJ C 139 (6 May 1995): 57.

60. Council Regulation no. 1210/90 of 7 May 1990 on the Establishment of the European Environment Agency and the European Environment Information and Observation Network, OJ L 120 (11 May 1990): 1.

61. "Delors Sees European Environment Agency, with Inspectors, but Timetable in Question," *International Environment Reporter* (14 June 1989): 287; "Parliament Calls for Stronger Role for Proposed European Environment Agency," *International Environment Reporter* (14 February 1990): 3.

62. House of Lords, Select Committee on the European Communities, HL Paper 53–I, *Implementation and Enforcement of Environmental Legislation,* vol. 1, *Report*, Session 1991–92, 9th Report (London: HMSO, 1992): 40–41.

63. Reflection Group's Report, 5 December 1995, 39; *Greening the Treaty II: Sustainable Development in a Democratic Union* (Utrecht, Netherlands: EEB, November 1995): 36; "Qualified Majority Vote in Council,

Enforcement by Court of Justice Proposed," *International Environment Reporter* (8 March 1995): 170–71.

64. There are, of course, parallels, for example, generally to the extent that both deal with the allocation of authority in multilevel political systems and specifically in the similarities between subsidiarity and popular sovereignty. See Scot Macdonald and Gunnar Nielsson, "Linkages between the Concepts of 'Subsidiarity' and Sovereignty: The New Debate over Allocation of Authority in the European Union." Paper presented at the Fourth Biennial Meeting of the European Community Studies Association, Charleston, S.C., May 1995. I argue simply that subsidiarity fundamentally differs from sovereignty on a number of important dimensions.

65. See Koen Lenaerts, "The Principle of Subsidiarity and the Environment in the European Union: Keeping the Balance of Federalism," *Fordham International Law Journal* 17 (1994): 846–95.

66. "Dictionary Time," *Economist* (9 December 1989): 52; "A Milder, Mellower Jacques Delors," *Economist* (27 June 1992): 56; Ludger Kuhnhardt, "Federalism and Subsidiarity," *Telos* 91 (Spring 1992): 77–86; Anthony L. Teasdale, "Subsidiarity in Post-Maastricht Europe," *Political Quarterly* 64 (April-June 1993): 187–97.

67. Kees van Kersbergen and Bertjan Verbeek, "The Politics of Subsidiarity in the European Union," *Journal of Common Market Studies* 32 (June 1994): 218.

68. John Pinder, "Economic and Social Powers of the European Union and the Member States: Subordinate or Coordinate Relationship," in Roland Bieber, Jean-Paul Jacqué, and Joseph H.H. Weiler, eds., *An Ever Closer Union: A Critical Analysis of the Draft Treaty Establishing the European Union* (Brussels: European Perspectives Series, 1985): 104–5.

69. Draft Treaty Establishing the European Union, OJ C 77 (19 March 1984): 38.

70. Draft Treaty, 47.

71. *Bull. EC* Supplement 2/86, Article 130r(4): 16.

72. *Treaty on European Union,* Article 3b, 13–14.

73. Andrew Duff, "Toward a Definition of Subsidiarity," in Duff, ed., *Subsidiarity within the European Community* (London: Federal Trust, 1993): 7–32.

74. Laurens Jan Brinkhorst, "Subsidiarity and European Community Environment Policy: A Panacea or a Pandora's Box?," *European Environmental Law Review* 2 (January 1993): 17; Paul Green, "Subsidiarity and European Union: Beyond the Ideological Impasse?," *Policy and Politics* 22 (October 1994): 287–300.

75. John Peterson, "Subsidiarity: A Definition to Suit Any Vision," *Parliamentary Affairs* 47 (January 1994): 118–19.

76. *Bull. EC* 10–1992, 116.

77. See *Bull. EC* 12–1992, 12–18; EP Doc. A3–0187/94, Report on the Commission Report to the European Council on the Adaptation of Community Legislation to the Subsidiarity Principle (Rapporteur: Medina Ortega, 29 March 1994): 6, 11.

78. Written question no. E–370/95 by Mr. Crampton, OJ C 175 (7 October 1995): 38. Golub contends that Britain was able to secure the withdrawal of several proposals to which it was opposed, and notes further that the number of environmental proposals offered by the commission declined significantly from 1992 to 1995. See Jonathan Golub, "Sovereignty and Subsidiarity in EU Environmental Policy," EUI Working Paper RSC no. 96/2 (Florence: European University Institute, 1996): 23.

79. David Gardner, "Brussels' Green Sprouts," *Financial Times* (21 October 1992): 18.

80. Draft Interinstitutional Agreement between the European Parliament, the Council, and the Commission on Procedures for implementing the Principle of Subsidiarity, OJ C 329 (6 December 1993): 135–36.

81. See, for example, the unnumbered commission document "Better Law-Making," Commission Report to the European Council on the application of the subsidiarity and proportionality principles, on simplification and on consolidation, 21 November 1995.

82. These difficulties are characteristic of so-called functional regimes; see Friedrich Kratochwil, "Of Systems, Boundaries, and Territoriality: An Inquiry into the Formation of the State System," *World Politics* 39 (October 1986): 48–50.

83. Caporaso, "The European Union and Forms of State," 47.

84. Mr. White, in *EP Debates* no. 3–426 (18 January 1993): 36.

85. Guenther F. Schaefer, "Institutional Choices: The Rise and Fall of Subsidiarity," *Futures* 23 (September 1991): 684–87.

86. Colin Hines, "The Green View on Subsidiarity," *Financial Times* (16 September 1992): 15; see also Santos, in *EP Debates* no. 3–426, 36.

87. "Psychic spillovers" are said to be present where there are noneconomic and nonphysical external costs to some activity. However, because almost any activity may have noneconomic and nonphysical external costs, using psychic spillovers as a criterion for locating authority would seem to conclude in favor of centralization in every instance. In economic terms, applying this idea would seem to place an infinite valuation on ecology. See Wouter P. J. Wils, "Subsidiarity and EC Environmental Policy: Taking People's Concerns Seriously," *Journal of Environmental Law* 6 (1994): 90–91.

88. Centre for Economic Policy Research, *Making Sense of Subsidiarity: How Much Centralization for Europe?* (London: CEPR, November 1993): 141–45.

89. Janne Harland Matláry, "New Forms of Governance in Europe? The Decline of the State as the Source of Political Legitimation," *Cooperation and Conflict* 30 (June 1995): 117.

90. EP Doc. A3–0380/92, Report of the Committee on the Environment, Public Health, and Consumer Protection on the Application of the Principle of Subsidiarity to Environment and Consumer Protection Policy (Rapporteur: Mr. White, 23 November 1992): 9.

91. See, for example, Regina S. Axelrod, "Subsidiarity and Environmental Policy in the European Community," *International Environmental Affairs* 6 (Spring 1994): 115–16.

92. Brinkhorst, "Panacea or a Pandora's Box?," 20.

93. COM(93) 545 final, Commission Report to the European Council on the Adaptation of Community Legislation to the Subsidiarity Principle (24 November 1993): 1.

94. Rodger A. Payne, "Freedom and the Environment," *Journal of Democracy* 6 (July 1995): 41–55.

95. Karlheinz Neunreither, "Subsidiarity as a Guiding Principle for European Community Activities," *Government and Opposition* 28 (Spring 1993): 209–10.

96. These terms are taken from James N. Rosenau, "Governance in the Twenty-First Century," *Global Governance* 1 (Winter 1995): 19–20.

97. Deborah Z. Cass, "The Word That Saves Maastricht? The Principle of Subsidiarity and the Division of Powers within the European Community," *Common Market Law Review* 29 (1992): 1126.

98. See Ulf Hedetoft, "The State of Sovereignty in Europe: Political Concept or Cultural Self-Image," in Staffan Zetterholm, ed., *National Cultures and European Integration: Exploratory Essays on Cultural Diversity and Common Policies* (Oxford: Berg, 1994): 13–48.

99. Brian Wynne, "Implementation of Greenhouse Gas Reductions in the European Community," *Global Environmental Change* 3 (March 1993): 128.

III

Revisioning Sovereignty

10

Eco-Cultural Security and Indigenous Self-Determination: Moving Toward a New Conception of Sovereignty

Sheldon Kamieniecki and Margaret Scully Granzeier

In this age of global environmental degradation and indigenous cultural insecurity, the shortcomings of conventional formulations of sovereignty are becoming increasingly apparent. Some scholars and practitioners are calling for a new conception of sovereignty—one that incorporates indigenous self-determination and recognizes the connection between environmental change and the expression of cultural rights.[1]

This chapter is but a first step in such a "reconceptualization." It begins with a brief overview and critique of conventional notions of sovereignty and then introduces the concept of *eco-cultural security* to challenge the traditional approach to the exercise of power.[2] Several case studies of indigenous cultural and environmental insecurity are provided to illustrate the benefits, drawbacks, and difficulties of enlarging the scope of sovereignty. The chapter concludes with some reflections on the potential effectiveness of a new concept of sovereignty in the context of the broad discretion and power currently exercised by states.

The Conceptual Foundations of Sovereignty

In the context of global environmental degradation and resource scarcities, sovereignty is generally thought to confer on states three specific spheres of legitimacy and power: (1) the ability to control territory and natural resources therein; (2) the right to exploit natural resources, and; (3) the authority to develop and enforce environmental regulations, standards, policies, and priorities in accordance with specific national interests and values.[3] Sovereignty is embedded in the concepts of national interest, national independence, and national security, and is commonly held to reflect "the notion of strength understood as the state's capacity to impose its will whether on its own citizens or other states."[4] Historically, sovereignty has been used to distinguish between order and anarchy, security and danger, and identity and difference.[5]

Sovereignty is typically equated with supreme power exercised within a particular territorial boundary.[6] As Camilleri and Falk observe, "Within national boundaries, the nation-state is supreme, recognizing no higher authority. Outside the national domain is the rest of the world, also partitioned into sovereign states which deal with each other, at least so far as their sovereignty is concerned, on a basis of equality."[7] In practice, sovereignty tends to cause assimilation and homogenization. Ostensibly, this is to ward off attacks from enemies, to maintain stability and peace, to foster an atmosphere in which growth and economic prosperity can occur, and to promote the formation of national pride and cultural identity. Indeed, sovereignty provides a label for a whole host of complicated, multifaceted relationships within and among states in the international system, allows for negotiation between otherwise disparate entities, and facilitates the standardization of rules and practices. As international legal scholar Malcolm Shaw explains, sovereignty "expresses internally the supremacy of the governmental institutions and externally the supremacy of the state as a legal person."[8]

While generations of scholars have employed this conceptualization of sovereignty to explain state behavior in the international system, this formulation is less helpful in cases involving indigenous, transnational, or nonstate actors in conflicts over the control and management of natural resources. The traditional international legal definition of sovereignty as constitutional independence does not further our understanding of competing forms of sovereignty (state and indigenous), is not sensitive to cultural variations, and may contribute to environmental degradation by hindering cooperation and indigenous self-determination.

Indigenous peoples in many countries, including the United States, find themselves caught in a precarious legal web of "semisovereign" status that complicates dispute resolution and accentuates the need to reevaluate the legal, political, and social standing of indigenous peoples globally. Furthermore, indigenous communities are frequently excluded from full political or social membership and participation and, accordingly, may not be afforded an adequate measure of protection, rights, or legitimacy. Political elites typically associate the term *self-determination* with secession and, inasmuch, may be threatened by indigenous demands for authority and control over traditional territories, natural resources, and cultural practices. In practice, law and policy may reflect fears about indigenous succession since states routinely discriminate against indigenous peoples, encroach upon their homelands, utilize their resources, and jeopardize their survival as a culture. The indigenous case raises sociocultural and political aspects integral to sovereignty that are not captured in timeworn, state-centric definitions of sovereignty.

Analyses of sovereignty must move beyond outdated models that answer demands for self-determination with perfunctory resistance and heightened fears about indigenous secession if scholarship is accurately to reflect contemporary global political and social life. Furthermore, scientific research on pollution and ecological change indicates that many serious environmental problems do not observe national boundaries, and may adversely affect the lives of citizens and alter the priorities and capabilities of states around the globe. Thus, attempts to determine cause-and-effect relationships and ensure cooperative efforts to mitigate environmental harm are frustrated by reliance on the doctrine of sovereignty, insulating states from the need to assume direct responsibility for actions detrimental to the health of the environment.

Grounding the notion of sovereignty in terms of territory, and by extension, title to property, has deep roots, but poses serious obstacles to contemporary efforts to protect indigenous self-determination, especially in light of diverse indigenous cultural worldviews. Frequently omitted from treatments of sovereignty and the global environment is the fact that attitudes about nature and its conservation (or exploitation) are culturally determined and culturally variant. The definitions of many ecological concepts, ranging from "territory," nature, and wilderness, to environmental risk, endangered species, and pollution, often differ widely across cultures, thus complicating our grasp of the sovereignty/environment linkage in multiple cultural contexts. This is not surprising, yet even a modest survey of international environmental law and policy reveals a tendency to universalize and to export Western and industrial assumptions, priorities, and regulatory strategies for environmental protection. A new formulation of sovereignty, therefore, must be culturally sensitive if it is to address the enormous environmental and human rights issues at hand.

Sovereignty and Sources of Environmental Degradation

The conventional doctrine of sovereignty privileges states' authority over their own resource management, thus allowing states to circumvent environmental regulations that do not coincide with the "national interest" or the objectives of economic development. Some scholars consider the state itself as the primary source of global environmental degradation and the main obstacle to effective environmental protection.[9] From this state-centric environmental perspective, sovereignty is considered an impediment to concerted international ecological protection, or an underlying cause of the world's environmental woes.

Other scholars point to the structure of the international system, and the inequitable distribution of wealth and power between and within industrialized and developing countries, as a source of the international environmental crisis.[10] Some investigators regard the lack of an overarching sovereign authority to inform, create, and enforce effective environmental regulations as ultimately responsible for hindering ecological protection.[11]

Many researchers emphasize the importance of decentralized political power and authority, and are especially interested in assessing the ability of transnational, nongovernmental, and local-level environmental movements, organizations, and nonstate actors to influence the outcome of environmental negotiations.[12] Still others underscore the important role of indigenous peoples, members of the so-called Fourth World, in the debate over environmental protection and sovereignty.[13] Empirical evidence confirms that the health of the environment is of critical importance to the continued survival, security, and general well-being of these communities.[14] This chapter emphasizes the core concept of self-determination, implicit in all discussions of sovereignty, and in this way facilitates the evaluation of indigenous natural resource management issues and cultural rights claims.

Environment, Culture, and Indigenous Peoples

Some indigenous rights advocates suggest that the use of "sovereignty" and "self-determination" metaphors perpetuates the historical marginalization of indigenous cultural, religious, and political rights.[15] Implicit in the definition of sovereignty, these authors contend, are the kernels of coercive control, manipulation, and assimilationist tendencies that are invoked by some states in order to survive external threats.[16] Other commentators note that the language of tribal sovereignty has been cleverly exploited by many governments seeking partnership with indigenous peoples in environmentally questionable projects, such as the storage and disposal of radioactive waste.[17]

On the one hand, full recognition of indigenous cultural and religious rights (i.e., indigenous sovereignty) may be a fundamental prerequisite to the elimination of the legacy of colonial imperialism and development in the name of "progress."[18] Indigenous self-determination, it is suggested, establishes an effective bulwark against continued discrimination and "eco-imperialism" or "eco-colonialism."[19] On the other hand, some investigators claim that the recent emphasis on the intersection of indigenous cultural rights and environmental protection reinforces state hegemony and reaffirms the difficulties inherent in addressing the globalization of new patterns of inequality and dominance.[20] In other words, the continued marriage by most states to the doctrine of sovereignty may indicate to indigenous peoples that economic development, industrialization, rapid growth, military prowess, and technological advancements will be sought and gained at the expense of environmental health, as well as cultural and religious rights.[21]

An alternative approach that bridges the gap between indigenous ideas, self-determination, various assertions of sovereignty, and environmental protection is that of "eco-cultural security." The notion of eco-cultural security suggests that severe environmental degradation or resource scarcities can endanger the continued survival of a particular group. Thus security is not limited to traditional military-strategic concerns, but includes a broader range of threats to environmental and human well-being and survival. If severe environmental harm compromises an indigenous group's fundamental cohesion, traditional customs, language, and homeland, then a security threat exists. As Vitit Muntarbhorn, a scholar in the field of environmental issues and indigenous rights, states:

> [Cultural] disintegration is compounded by destruction of the ecology and habitat upon which indigenous groups depend for their physical and cultural survival. Deforestation, particularly of rain forests, and pollution induced by outsiders, jeopardize the modus vivendi of indigenous groups. The social nexus binding members of the group to the environment is thus annihilated.[22]

Threats to cultural and environmental integrity may include large-scale mining, logging, agricultural conversion, hydroelectric or geothermal development projects, or natural disasters (the effects of which may be exacerbated by human-induced environmental change).

Conventional wisdom posits that principles enshrined in sovereignty and in the process of state building, including the accumulation of wealth, development, and security, are considered by political elites to be more important than the concerns of indigenous or cultural minorities. Additionally, the structure and distribution of power in the international system impede the ability of indigenous communities to prevent and respond to much of the environmental harm that threatens their existence. The simultaneous abuse of indigenous rights and exploitation of natural resources is found in most nations around the world. Anthropologists have long observed the close relationship between severe environmental degradation and incidence of disease, conflict, sociocultural and economic breakdown, and refugeeism in indigenous communities.

Scholars are calling for a reconsideration of sovereignty in light of mounting environmental threats. These threats may require nation-states to relinquish a certain degree of control and authority over their use, management, and protection of natural resources. Otherwise, sovereignty is an impediment to constructive, cooperative international efforts to protect the natural environment. Analyses of new approaches to sovereignty should incorporate indigenous self-determination and threats to eco-cultural security to enrich and compliment the traditional investigation of state actors, political institutions, and regime formation.

Threats to Eco-Cultural Security

There are numerous examples of how the effects of environmental disruption and human rights violations threaten the survival of indigenous peoples. These cases arise in countries with varying democratic traditions and demonstrate the intrinsic relationship between several factors: severe environmental damage, human rights violations, attempts to preserve cultural and religious heritage, and the potential for insecurity and marginalized claims to indigenous sovereignty.[23] As the special rapporteur of the United Nations Subcommittee on Prevention of Discrimination and Protection of Minorities observes,

The denial of the right of peoples to self-determination and practices . . . characterized by massive and systematic violations of human rights, lie[s] at the origin of the degradation suffered by the environment in those territories and of the damage done to the cultural heritage and living conditions of the population, who are reduced to living in camps, precarious housing, and areas bereft of basic sanitation.[24]

Several examples of indigenous eco-cultural insecurity are briefly described here to illustrate the importance of incorporating nonstate actors and environmental factors into analyses of sovereignty and security. In addition, these examples demonstrate the various ways that sovereignty is employed in disputes over economic development, cultural rights, and environmental protection: as a rhetorical device, a tool for manipulating less powerful actors, or as a partial justification for unpopular political, social, or economic decisions. Indeed, sovereignty has many faces; it can be invoked to protect the environment, to exploit the environment, to protect indigenous rights, and to prevent the full exercise of those same rights.

The Yanomami in Brazil

A stark example of the combined effects of cultural and environmental damage on an indigenous community's survival is the case of the Yanomami Indians in Brazil. In this situation, indigenous eco-cultural insecurity has undermined the expression of self-determination and even led to loss of life. In 1973 the Brazilian government began the construction of the Perimetral Norte Highway along the northern national border.[25] Anthropological records indicate that during the construction the Yanomami were "in a state of misery, sickness, and shock. They refused to speak their language, their gardens had been uprooted by bulldozers, and they wore ragged clothing given to them by the construction workers, which was infested with influenza, tuberculosis, measles, and other germs."[26] The discovery of uranium cassiterite and other valuable minerals in Yanomami territory in the early 1980s brought more than 40,000 prospectors, several airstrips, establishment of mining camps, and a host of foreign diseases to traditional Yanomami lands.[27] Some 1,500 people, an estimated 15 percent of the Brazilian Yanomami population, died as a result of violence, malnutrition, malaria, and other diseases introduced to the area by gold prospectors and other miners.[28]

Several international nongovernmental human rights and environmental organizations monitoring the Yanomami case submitted a complaint to the Inter-American Commission on Human Rights on behalf of the Yanomami.[29] In this unprecedented case, the commission found that environmental degradation, precipitated by development activities and in tandem with natural factors, can indeed threaten cultural security and violate the right to life.[30] In particular, the commission found that the Brazilian government's failure to take timely and effective measures to prevent environmental harm leading to the loss of cultural identity, property, and life had constituted grave violations of human and cultural rights.[31]

The commission stated that international law currently "recognizes the right of ethnic groups to special protection on the use of their own language, for the practice of their own religion, and, in general, for all those characteristics necessary for the preservation of their cultural identity."[32] Although this case afforded a symbolic victory to the Yanomami, logging and mining activities continue to jeopardize the security and survival of these and other indigenous peoples.[33] In fact, in 1993, more than 11,000 miners and prospectors were thought to be operating illegally in Yanomami territory. The historical legacy of considering lands occupied by indigenous peoples as *terra nullius*, or empty land, persists.[34]

The Penan in Malaysia

The fate of another indigenous community, the Penan in Malaysia, demonstrates how powerful states use sovereignty to justify unsustainable environmental practices and policies that endanger the cultural survival of the minority communities living within their borders. In Malaysia, enormous logging operations and forestry ventures, responsible for one of the highest rates of deforestation in the world, have denuded one-third of the forested land in the state of Sarawak in just two decades.[35] The Malaysian government acknowledges that pollution, soil erosion, and loss of aquatic and wildlife ecosystems are important environmental and epidemiological concerns.[36] In practice, however, intense pressure to industrialize and compete in the global marketplace, combined with domestic political factors (lucrative logging concessions are owned by political elites in Malaysia) serve to destroy the environment and compromise the security and existence of the Penan people.

In addition, the absence of democratic practices and institutions in Malaysia has made it extremely difficult and even dangerous for the Penan to oppose political and corporate interests or to express their rights to indigenous self-determination and control over traditional homelands and natural resources.[37] The Penan have been forcibly evicted from their homelands and have been imprisoned for expressing opposition against the activities of logging concessions.[38]

The Penan have a spiritual and physical dependence on nature that is linked to their subsistence practices and religious tenets. As the result of deforestation, the Penan will lose many species of plants used for medicinal, religious, and artistic purposes and may have to abandon their traditional relationship with the environment.

What if there had existed a widely accepted, alternative conception of sovereignty that recognized the importance of eco-cultural security and the right to self-determination of indigenous communities? Would the Yanomami and Penan have been spared? This question illustrates the limitations of any definitional change: The structure of the international system and the distribution of power within it, the globalization of capitalism, and the concomitant pressures to develop all militate against environmental and cultural protection. In order for a new conception of sovereignty to be effective, it would have to incorporate ethical considerations of state restraint and respect for human dignity and environmental health.

Native Americans in the United States

A third example of the importance of eco-cultural security to indigenous groups is the case of Native Americans in the United States.[39] Despite strong support in the United States for human rights and democratic ideals, historical patterns of the abuse, marginalization, and rejection of various Native American cultural and religious rights has been well documented. In the United States, courts have historically rejected Native American cultural and religious rights claims on the grounds that: (1) no undue infringement upon Native American religions existed; and (2) the proposed infringement on the sacred site was justified by "countervailing state interests."[40] Complicated questions regarding Native American cultural rights to use endangered species, such as bald eagles or black panthers, in ceremonial practices; the acceptance of nuclear waste on tribal lands; the preservation of sacred sites in the face of development; and self-determination and sovereignty are contentious issues receiving attention in anthropology, environmental law, ethics, and geography.

Native American tribes frequently invoke the language of sovereignty in disputes over particular territory or natural resources.[41] Native Americans stand to lose many lands to modern development as more than three hundred sacred sites are currently under consideration for development.[42] For example, Mount Graham, the home of the sacred Crown Dancers for the San Carlos Apache in Arizona, is the proposed site of a multimillion dollar astronomical observatory named Columbus Project.[43] Near Mount Shasta, California, both uranium miners and a proposed new ski resort threaten Panther Meadows, an area sacred to the Wintu.[44]

In 1988 the United States Supreme Court, in *Lyng v. Northwest Indian Cemetery Protection Association*,[45] upheld a U.S. Forest Service plan to construct a logging road through the Chimney Rock area of the Six Rivers National Forest in California, an area of "high country" sacred to several California tribes. The court conceded that the "threat to the religion was extremely grave," and it acknowledged that the road would "virtually destroy the Indians' ability to practice their religion."[46] Nevertheless, the court held that this destruction of cultural traditional practices was not an undue burden to the Native Americans under the free exercise clause of the United States Constitution.[47] In the court's opinion, cultural rights to control ancestral homelands and preserve a sacred burial ground were not sufficient to restrict the economic and strategic interests of the U.S. government. The court reasoned that the construction of the highway and logging activities in the area did not constitute punishment or coercion of Native Americans for practicing their religion, and therefore, the government did not have to restrict development and resource extraction in the area.

In making their respective arguments in this case, the federal government and the Native American governments both asserted sovereign rights. A ruling for indigenous sovereignty over resources would have led to the preservation of culture and the environment since the Apache, Yurok, Karok, and Tolowa tribes in the region would have protected the sacred burial ground. Usually, however, the exercise of sovereignty results in the degradation of natural resources, at least where nation-states in the international system are concerned.

Another example of the relationship between the twin issues of environmental protection and cultural survival, and Native American claims of sovereignty, involves the disposal and storage of solid and hazardous waste on reservation land of the Campo tribe in California. It illustrates the complicated interplay of diverse interests in democratic societies. The Campo tribal government has sought to participate in the construction of a large solid waste treatment and storage facility on the Campo reservation in return for significant economic gain. The tribal council, after reviewing the costs and benefits of such a plan, asserted its rights to self-determination and, consequently, its right to pursue economic development. In response, many environmentalists, other tribal nations, and concerned advocates on Native American issues have lobbied strenuously against the proposed landfill site, and have taken the matter to court in California.

Many Campos cited by Dan McGovern in his account of the dispute suggested that opposition to this development stems from long-standing paternalism and discrimination.[48] Others described a dilemma between respect for nature and environmental concerns central to Native American cultural worldviews, on the one hand, and the need for income to survive as a culture, on the other. Although no consensus exists within the Campo tribal government, the proponents of the solid and hazardous waste facility consider the project a mechanism through which to combat their sense of powerlessness, to participate on equal terms with the U.S. government, and to make decisions that effect their future.[49]

Opponents of the landfill construction suggest that the U.S. Department of Energy (DOE) is using the rhetoric and promise of "tribal sovereignty" as a valuable commodity. Critics point to the lucrative "grants" offered by the DOE to tribal leaders willing to take radioactive waste from nuclear power plants and medical waste in the name of "Native American sovereignty." Some scholars suggest that, by entering into "ill-advised resource extraction of hazardous waste disposal schemes," Native American groups, like the Campo Indians, are subjecting themselves to the "market recolonization" of their traditional homeland and cultural practices, as "prodevelopment interests have fixed on the economic plight of indigenous peoples as a rationale for enabling and urging them to engage in maximal resource exploitation."[50]

This case reveals a fundamental limitation in the possibility of achieving environmental protection through the reconsideration of sovereignty. Environmentalists may advocate the full recognition of tribal sovereignty, but may also be troubled if the expression of sovereignty does not correspond with romantic depictions of traditional societies. Clearly, there is no guarantee that reconceptualizing sovereignty to include the self-determination of indigenous peoples will automatically lead to the preservation of culture and the protection of the environment in every case.

Some Final Thoughts

This chapter has shown how traditional formulations of sovereignty, by excluding nonstate actors, provide an inherently incomplete picture of world politics. While investigators are increasingly cognizant of the difficulties presented to nation-states by environmental crises, the relationships between environmental degradation, economic development activities, and the cultural and human rights of indigenous people have not been adequately explored.

The term *eco-cultural security* is employed to describe a cluster of threats to indigenous groups, including severe ecological harm, human and cultural rights violations, and the combined effects of poverty, discrimination, and disenfranchisement on cultural survival and political participation. More research is necessary to show how global patterns of environmental injustice might be related to indigenous demands for sovereign recognition, land rights, and self-determination.

Full recognition of indigenous sovereignty and protection of cultural rights, while extremely compelling on moral and philosophical grounds, is unlikely to occur in the near future. Imagine an international system in which cultural minorities recovered ancestral territory; enjoyed a wide spectrum of rights, freedom, and respect; and engaged in cultural practices without fear of censure. Would indigenous communities survive the transition and the onslaught of nonindigenous worldviews with their traditional cultures intact? Does indigenous sovereignty preclude cultural transformation, or does it require such a transformation? It is not clear whether, as Corntassel and Primeau contend, indigenous peoples' calls for self-determination or for absolute rights to self-identification would exacerbate tensions between groups and states, and ultimately result in intractability and outright hostility on the part of host states.[51] Powerful states would stand to lose a great deal in this new system, and there are few incentives to induce them to approve such radical restructuring of law and policy. Continued efforts by grass roots activists and other concerned citizens may influence gradually the creation of laws and policies that strive toward cross-cultural accommodation, although cultures are not static and certain changes may be inevitable.

The implication of environmental degradation (and corresponding regulatory strategies designed to protect nature) for indigenous sovereignty is another area requiring further investigation. For example, if the eco-cultural security of indigenous cultures is jeopardized due to environmental factors (or plans for environmental protection), then what measures are (or ought to be) in place to protect the sovereignty of these indigenous communities? Alternatively, if the industrialized world is responsible for the environmental, social, political, and economic harm experienced by indigenous groups worldwide, then are efforts to mitigate this damage a contemporary form of paternalism? Is it possible for states to accept responsibility for global environmental harm, and the negative consequences it has had on particularly vulnerable societies, without further compromising an indigenous group's ability to maintain its identity and control its own destiny?

This chapter does not suggest that traditional democratic practices, procedures, and institutions commonly found in the West must be adopted by all nations in order to conserve natural resources. Generally, democratic principles facilitate the adoption of pollution control regulations and natural resource conservation programs; however, they are often not present in nations in which so many indigenous people find themselves. Even then, there is no assurance that if democratic principles were adopted and implemented by governments of countries with indigenous communities that the predicament of indigenous peoples would improve, as the case of Native Americans in the United States demonstrates. Moreover, many principles of democratic theory may not apply to indigenous cultures, given their varied social, legal, and political ideas and arrangements.

Clearly, the right to self-determination for indigenous peoples is far more desirable on moral grounds than the continued repression, exploitation, and eventual demise of these peoples. Changes in values, knowledge, and common practices inside and outside indigenous communities are necessary if cultures are to be protected and effective conservation measures are to be implemented. Such changes are most likely to occur in free and open democratic societies, yet they are most needed in less democratic societies.

This chapter has argued that attacks against the cultural and environmental rights of indigenous peoples have undermined their rights to self-determination and have threatened their existence. The traditional notion of sovereignty permits and often exacerbates threats against their eco-cultural security. Thus, any reconceptualization of sovereignty that attempts to respond to the changing social, economic, and political dynamics in the world must include the right to self-determination of indigenous groups. While indigenous peoples will not always make decisions that are popular or supported by environmentalists, they will at least be given the opportunity to shape their own destinies. Future studies should therefore attempt to develop new definitions of sovereignty that incorporate eco-cultural security considerations of indigenous communities so that quality of life is enhanced and the chances for indigenous peoples' survival are increased.

Notes

1. See Robert H. Jackson, *Quasi-States: Sovereignty, International Relations, and the Third World* (New York: Cambridge University Press, 1990); Joseph A. Camilleri and Jim Falk, *The End of Sovereignty? The Politics of a Shrinking and Fragmenting World* (Aldershot, England: Edward Elgar Publishing, 1992); Ruth Lapidoth, "Sovereignty in Transition," *Journal of International Affairs* 45, no. 2 (Winter 1994): 325–46; Sanford Lakoff, "Between Either/Or and More or Less: Sovereignty Versus Autonomy under Federalism," *Publius* 24, 1 (Winter 1994): 63–78; and Ole Wæver, "Identity, Integration, and Security: Solving the Sovereignty Puzzle in EU Studies," *Journal of International Affairs* 48.

2. Margaret Scully Granzeier, "Linking Environment, Culture, and Security" in Sheldon Kamieniecki et al., eds., *Flashpoints in Environmental Policy Making: Controversies in Achieving Sustainability* (Albany, State University of New York Press, 1997).

3. These principles of sovereignty are affirmed in Principle 21 of the 1972 Report of the United Nations Conference on the Human Environment, or "Stockholm Declaration": "States have, in accordance with the Charter of the United Nations and the principles of international law, the sovereign right to exploit their own natural resources pursuant to their own environmental policies, and the responsibility to ensure that activities within their jurisdiction or control do not cause damage to the environment of other States or of areas beyond the limits of national jurisdiction" (United Nations Stockholm Declaration, 1972). See Stanley M. Spracker, "Sovereignty and the Regulation of International Business in the Environmental Area: An American Viewpoint," *Canada–United States Law Journal* 20 (1994): 225–35; and Janice E. Thomson, "State Sovereignty and International Relations: Bridging the Gap between Theory and Empirical Research," *International Studies Quarterly* 39 (Summer 1995).

4. Camilleri and Falk, *The End of Sovereignty,* 11.

5. Although the realist tradition in international relations theory treats the principle of state sovereignty as an absolute, fixed, unitary construct, critics of this paradigm have observed the transitional nature of sovereignty and its multidimensional overtones. See Camilleri and Falk *The End of Sovereignty*; J. Samuel Barkin and Bruce Cronin, "The State and the Nation: Changing Norms and the Rules of Sovereignty in International Relations," *International Organization* 48, no. 1 (Winter 1994): 107–30; Franke Wilmer, *The Indigenous Voice in World Politics: Since Time Immemorial* (Newbury Park and London: Sage Publications, 1993); and Paul Wapner, "The State and Environmental Challenges: A Critical Exploration of Alternatives to the State System," *Environmental Politics* 4, no. 1 (1995): 44–69.

6. Janice Thomson, "State Sovereignty in International Relations," 213–33.

7. Camilleri and Falk, *The End of Sovereignty,* 3.

8. Malcolm N. Shaw, *International Law*, 2d ed. (Cambridge: Grotius Publications, 1986): 238.

9. Barry Buzan, *People, States, and Fear: An Agenda for International Security Studies in the Post–Cold War Era*, 2d ed. (Boulder, Colo.: Lynne Rienner, 1991); Camilleri and Falk, *The End of Sovereignty*: and Andrew Hurrell, "A Crisis of Ecological Viability? Global Environmental Change and the Nation-State," *Political Studies* 42 (1994): 146–65.

10. Naeem Inayatullah and David L. Blaney, "Realizing Sovereignty," *Review of International Studies* 21 (1995): 3–20.

11. Lester Milbrath, *Envisioning a Sustainable Society: Learning Our Way Out* (New York: State University of New York Press, 1989); Richard Falk, "Toward a New World Order Respectful of the Global Ecosystem," *Boston College Environmental Affairs Law Review* 19, no. 4 (1992): 711–24; and Ronnie Lipschutz and Ken Conca, eds., *The State and Social Power in Global Environmental Politics* (New York: Columbia University Press, 1993).

12. Wapner, "The State and Environmental Challenges."

13. Wilmer, *The Indigenous Voice in World Politics*.

14. Jason Clay, "Looking Back to Go Forward: Predicting and Preventing Human Rights Violations," in Marc S. Miller, ed., *State of the Peoples: A Global Human Rights Report on Societies in Danger* (Boston: Beacon Press, 1993); Art Davidson, *Endangered Peoples* (San Francisco: Sierra Club Books, 1993); and J. Baird Callicott, *Earth's Insights: A Survey of Ecological Ethics from the Mediterranean Basin to the Australian Outback* (Berkeley: University of California Press, 1994).

15. Jeff J. Corntassel and Tomas Hopkins Primeau, "Indigenous 'Sovereignty' and International Law: Revised Strategies for Pursuing 'Self-Determination'" *Human Rights Quarterly* 17 (Spring 1995): 343–65.

16. Leo Kuper, *Genocide: Its Political Use in the Twentieth Century* (New Haven and London: Yale University Press, 1981); Wilmer, *Indigenous Voice in World Politics*; and Corntassel and Primeau, "Indigenous 'Sovereignty' and International Law."

17. Charles K. Johnson, "A Sovereignty of Convenience: Native American Sovereignty and the United States Government's Plan for Radioactive Waste on Indian Land," *St. John's Journal of Legal Commentary* 9, no. 2 (Spring 1994): 589–97.

18. Wilmer, *Indigenous Voice*; and Lloyd Burton, "Indigenous Peoples and Environmental Management in the Common Law Nation-States of the Pacific Rim: Eco-Colonialism or Market Recolonization?" Paper presented at the annual meeting of the Law and Society Association, April 1995, Toronto, Canada.

19. Burton, "Indigenous Peoples and Environmental Management."

20. Hurrell, "A Crisis of Ecological Viability?"; and Wapner, "The State and Environmental Challenges."

21. Robert Hitchcock, "International Human Rights, the Environment, and Indigenous Peoples," *Colorado Journal of International Environmental Law and Politics* 5, no. 1 (Winter 1994): 1–22; and Dean B. Suagee and Christopher T. Stearns, "Indigenous Self-Government, Environmental Protection, and the Consent of the Governed: A Tribal Environmental Review Process," *Colorado Journal of International Environmental Law and Politics* 5, no. 1 (Winter 1994): 23–59.

22. Vitit Muntarbhorn (background report) E/CN.4/1989/22, annex III C, para. 3. Cited in Fatma Ksentini, "Human Rights and the Environment: Final Report," prepared by Fatma Ksentini, Special Rapporteur, Commission on Human Rights, Subcommittee on Prevention of Discrimination and Protection of Minorities. E/CN.4/sub.2/1994/9. 6 July 1994.

23. The special rapporteur of the Commission on Human Rights, Subcommittee on Prevention of Discrimination and Protection of Minorities, notes that indigenous "human rights violations ... almost always arise as a result of land rights violations and environmental degradation and indeed are inseparable from these factors" (Ksentini, "Human Rights and the Environment," para. 88).

24. Ksentini, "Human Rights and the Environment," para. 164.

25. Leslie Sponsel, "The Yanomami Holocaust Continues," in Barbara Rose Johnston, ed., *Who Pays the Price? The Sociocultural Context of Environmental Crisis* (Washington, D.C.: Island Press, 1994): 37–47.

26. Davis, "Land Rights and Indigenous Peoples,"

27. Davis, ibid.; and Sponsel, "The Yanomami Holocaust Continues."

28. *Yanomami Case*, Case 7615, March 5, 1985, in the annual report of the Inter-American Commission on Human Rights, 1984–1985 (OEA/Ser.L/V/II.66, doc.10, rev. 1).

29. Ibid.

30. Ibid.

31. Ibid.

32. Ibid.

33. Davis, "Land Rights"; Clay, "Looking Back to Go Forward"; and Davidson, *Endangered Peoples.*

34. Hitchock, "International Human Rights."

35. Davis, "Land Rights."

36. Ibid.

37. Ibid.; Hitchcock, "International Human Rights."

38. Ibid.

39. It is necessary to point out here that, although Native Americans are quite diverse, there are enough similarities in terms of conceptualizations of nature, among other cultural and religious traditions, and with regard to their treatment historically by the U.S. government, to warrant such a cross-cultural examination of Native American issues.

40. This centrality and sincerity threshold test does not take into consideration that most Native Americans do not have "an authoritative doctrine that sets a hierarchy of religious importance." Furthermore, this test uses a framework of analysis that may not be appropriate in the context of non-Western cultures. See Richard Herz, "Legal Protection for Indigenous Cultures: Sacred Sites and Communal Rights," *Virginia Law Review* 79, no. 3 (1993): 705.

41. S. James Anaya, "The Native Hawaiian People and International Human Rights Law: Toward a Remedy for Past and Continuing Wrongs," *Georgia Law Review* 28, no. 2 (1994): 309–65; Calicott, *Earth's Insights*; Jenny Honz, "Sacred Sites, Disputed Rights," *Human Rights* 19, no. 4 (1992): 26–29; Suagee and Stearns, "Indigenous Self-Government."

42. Sacred sites may be significant to a Native American group for a variety of reasons — historical, ceremonial, religious, etc. — although the particular importance of sacred sites varies by group. See Herz, "Legal Protection for Indigenous Cultures," 691–717. Walker has compiled an extensive list of more than three hundred. Native American sacred sites and their significance. See Deward E. Walker, "Protection of American Indian Sacred Geography," in Christopher Vecsey, ed., *Handbook of American Indian Religious Freedom* (Crossroad, N.Y., 1991).

43. The irony of the name of the observatory has not been lost on Native Americans, environmentalists, and their advocates. See Honz, "Sacred Sites, Disputed Rights," 26–29.

44. Herz, "Legal Protection for Indigenous Cultures."

45. *Lyng v. Northwest Cemetery Association*, 1988, 485 US 439.

46. Ibid.

47. Ibid.

48. Dan McGovern, *The Campo Indian Landfill War: The Fight for Gold in California's Garbage* (Norman: University of Oklahoma Press. 1995).

49. McGovern, *Campo Indian Landfill War.*

50. Burton, "Indigenous Peoples and Environmental Management," 13.

51. Corntassel and Primeau, "Indigenous 'Sovereignty' and International Law."

11

Reorienting State Sovereignty: Rights and Responsibilities in the Environmental Age

Paul Wapner

Hemisphere solidarity is new among statesmen, but not among the feathered navies of the sky.
—Aldo Leopold[1]

Environmental issues pose new challenges for state sovereignty, which is traditionally defined as political, military, administrative, and juridical control over a given territory. Sovereignty grants states ultimate authority over their own affairs, including choices about how to treat the natural world within their domain. However, contemporary environmental threats, from ozone depletion and global warming to polluted waters and eroding soils, transcend state boundaries and thus demand actions that encroach upon state sovereignty. States are being asked to curtail or alter their development and environmental policies to harmonize with those of their neighbors and, in some instances, the entire international community.

Since the early part of this century, with the emergence of transboundary environmental problems, states have wrestled with the issue of sovereignty as it relates to environmental issues. This has involved trying to balance a state's right to pursue policies within its domain against the responsibility to ensure that such policies do not adversely affect others. Regimes involving the renegotiation of the relationship between sovereignty and environment have been created through bilateral and multilateral action plans, conventions, declarations and institutional mechanisms, and conferences such as the United Nations Conference on Environment and Development (UNCED). Notwithstanding the significance of these efforts, sovereignty remains problematic. Capturing the appropriate balance between rights and responsibilities is no easy matter, especially when the issues are as complex as the interface between social and ecological systems, and when the pace of change within these systems is continuously accelerating due to increases in human activity.

This chapter analyzes the tension between rights and responsibilities as it finds expression at the intersection of what Litfin calls the "ecology/sovereignty nexus." I do not seek to resolve the tension so much as to understand it, arguing that the ecology/sovereignty nexus can best be understood as a struggle to marry rights and responsibilities. The chapter additionally has a normative agenda in that I assume that an increased emphasis on responsibility rather than rights would serve the environment well; it would diminish the self-regarding orientation states tend to assume in international fora and orient them toward a commitment to preserve the earth's biophysical well-being. The normative dimension of the chapter consists, then, in prescribing a number of ways of enhancing global environmental responsibility in the context of the current state system. Overall, I hope to render a greater appreciation for the problematic character of state sovereignty with regard to environmental issues and to recommend a number of policy prescriptions for moving beyond it.

Sovereignty and Global Environmental Protection

As other contributors to this volume make clear, sovereignty is difficult to define. Its meaning changes as one approaches it from legal, sociological, historical, or practical perspectives.[2] For present purposes, *sovereignty* refers simply to a combination of authority and control that states enjoy over domestic affairs within their territorial domains. Such authority and control comes, on the one hand, from mutual recognition of juridical rights among states and, on the other, from a state's capacity to shape collective affairs within its territory — a capacity that varies tremendously among different states.[3] In environmental affairs, the notion of sovereignty translates into both the entitlement and the heterogeneous ability of states to pursue environmental and developmental policies within their own territories as they see fit.

The territorial dimension stands at the center of the ecology/sovereignty nexus. While sovereignty suggests that states have authority and control over their own territories, those territories themselves are part and parcel of the global ecosystem and cannot be isolated in any meaningful fashion. Ecological interdependence thus scrambles the logic of sovereignty as it relates to environmental issues. This is particularly true of shared resources—biophysical systems such as regional seas, river beds, stocks of migratory animals, and so forth that span national territories—and issues of the global commons such as stratospheric ozone depletion and climate change. The meaning of sovereignty is also problematic in cases of transboundary externalities, instances where activities within one state's jurisdiction adversely affect conditions within another state or states.[4] In all three cases, environmental dynamics transcend national boundaries and thus problematize the allocation of political authority and control. The challenge for states is how to appropriate understandings and practices that reconcile the seeming mismatch between territoriality and ecological protection. Understandably, this challenge has turned on the issue of sovereignty.

The concept of sovereignty has gone through a long but ultimately slow evolution with regard to international environmental issues.[5] Although numerous doctrines have addressed the issue of sovereignty, the first to focus specifically on environmental issues and sovereign authority, and the one that set the stage for much subsequent thinking about the relationship, was the Harmon Doctrine. In 1895, U.S. Attorney General Judson Harmon articulated sovereignty's initial environmental meaning when he dismissed Mexican complaints that U.S. use of the Rio Grande adversely affected Mexico. According to Harmon, the United States had the legal right to use waters within its territory as it saw fit, independent of external effects. The so-called Harmon Doctrine was based on the view that, "The fundamental principle of international law is the absolute sovereignty of every nation, as against all others, within its own territory."[6] The effort was, obviously, to underscore the concept of rights rather than responsibility—to emphasize entitlement, as it relates to sovereignty, rather than obligation, as it relates to transboundary processes or activities. Subsequent cases and advancements in legal thought and practice, especially relating to air pollution and riparian rights theory, moderated Harmon's absolute notion of sovereignty and required that states use resources within their territories in ways that do not damage adjoining states or coriparians. This notion, which was codified in the doctrine *sic utere tuo ut alienum non laedas* (one must use his own so as not to injure others) and supported by the general principle of good neighborliness stands at the heart of contemporary conceptions of sovereignty and environmental protection.[7] It seeks a balance between rights and responsibilities, reminding states that, while they have authority over their own territories, ecosystems transcend political boundaries and extend obligations. Sovereignty, thus conceptualized, is "restricted."[8] Resources within one's boundaries are interconnected with environmental systems in other states, and thus states must pay heed to the extraterritorial ramifications of activities within their borders and, it is important to point out, their activities in global common areas.[9]

The enshrinement of this notion is captured most elegantly and forcefully in the Stockholm and Rio declarations, which emanated respectively from the 1972 United Nations Conference on Human Environment (UNCHE) and the 1992 United Nations Conference on Environment and Development (UNCED).[10] Principle 21 of the Stockholm Declaration proclaims that

States have, in accordance with the Charter of the United Nations and the principles of international law, the sovereign right to exploit their own resources pursuant to their own environmental policies, and the responsibility to ensure that activities within their jurisdiction of control do not cause damage to the environment of other States or of areas beyond the limits of national jurisdiction.[11]

This formulation, which some call a "triumph of diplomatic compromise,"[12] is woven throughout the 1972 Action Plan for the Human Environment insofar as many recommendations specify instances in which environmental abuse in one country will adversely affect environmental quality in another or where states share contiguous protected areas.[13] Moreover, its encapsulation is so powerful that twenty years later, at UNCED, states reiterated the formulation almost verbatim in Principle 2 of the Rio Declaration.[14] Principle 2 reaffirms that states have the right to pursue resource policies within their domains without outside interference and, in so doing, that they must consider the external effects on other states.

Global Responsibility and State Mechanisms of Governance

The effort to link rights and responsibilities testifies, at least at the general level, to a common understanding about the problematic nature of state sovereignty as it relates to environmental challenges. Moreover, it reveals a stated normative commitment to restrict the freedom of individual states when one state's actions adversely affect another's. For all its impressiveness, however, translating such understanding and commitment into practice has proved difficult. The reason for this is that the "rights" side of the equation is much easier to understand, argue for, and practice than the "responsibility" side. States have a tremendous amount of experience upholding sovereignty in the international sphere; there is a long and established tradition of respecting state sovereignty and, indeed, it is one of the most entrenched norms of international life. On the other hand, there is much less experience with and support for the notion of responsibility. While states have always been partially mindful of how their activities affect others, mechanisms of governance that aim to concretize this mindfulness are relatively new and weak, and lack the support of self-regarding action. Responsibility is an afterthought, as it were, of international behavior. Principles 21 of the Stockholm Declaration and 2 of the Rio Declaration, then, might more accurately be called "triumphs of aspiration" in a tradition of discourse and practice animated primarily by upholding state sovereign authority. Rights rather than responsibilities are the known quantities of international relations.

The imbalance between prerogative and obligation was poignantly clear at UNCED. Oil-producing states, for example, rejected all references to petroleum as environmentally harmful. The United Arab Emirates, for instance, claimed that oil is a "clean" resource because it produces less carbon dioxide than coal.[15] Malaysia, the Philippines, and Brazil, key timber-producing states, opposed any regulations that would have curtailed their right to cut their forests at current, unsustainable rates. Their opposition contributed to downgrading a proposed forest treaty to a set of nonbinding principles.[16] The United States, for its part, opposed all references to lifestyle changes and worked to minimize references to the link between consumption and environmental destruction in Rio documents. At Rio, President George Bush declared that he was "president of the United States, not president of the world and [that he would] do what [was] best to defend U.S. interests."[17] These instances confirm the general insight that states, of their own accord, will not easily assume responsibility for the ecological well-being of the globe. They will first consider their own interests; global interests will often be advanced only when they clearly coincide with national ones.

Liability Regimes

The international community has, nonetheless, devised mechanisms of governance to institutionalize global responsibility. Of the many that exist, the most relevant with regard to environmental issues and the ones that directly engage the self-regarding character of the state system fit under the general categories of liability or regulatory regimes. Liability regimes involve specifying culpability in the event of extraterritorial damage. Specifically, they are relevant for cases in which individual states are to blame for damaging other states' ecosystems. Based often on the customary practices of notification, consultation, and duty-to-warn in cases of imminent threats, liability regimes work, in their best formulation, after harmful effects are experienced and culpability can be assigned. Impaired states bring complaints directly to those suspected of causing environmental damage — in the hopes of stopping offending practices or seeking compensation — or they frame their concerns into an aggrieved format and pursue litigation with third-party arbitration. An example of the former is when Mexico protested odors from a Texas stockyard in 1961, forcing the United States to apologize and alleviate the problem.[18] An example of the latter is the famous Trail Smelter case, in which Canada and the United States agreed to arbitration to redress the problems of fumes drifting into the state of Washington from a Canadian lead smelter. An international tribunal ruled in favor of the U.S. complaint and awarded damages, and called on Canada to prevent future smelter activities from harming U.S. citizens.[19] The idea behind liability regimes is that, in instances when customary principles of international law are violated, especially "good neighborliness," culpability and reparations can be determined. In other words, environmental degradation can be traced back to its source and the source must assume responsibility for its actions. The liability regime's strength, as a form of governance, lies in addressing single instances of transboundary environmental harm where hazardous activities clearly endanger identifiable victims.

While liability regimes provide concrete ways to extend responsibility, in practice they have been invoked in only a fraction of the cases of transboundary environmental harm, and pursuing diplomacy and litigation in such cases has been an immense challenge. One reason for this is that environmental liability regimes rest, almost exclusively, on a set of general international legal principles that are not specifically fashioned to address environmental issues. As mentioned, liability becomes an issue when the general doctrine of *sic utere tuo ut alienum non laedas* has been violated. Indeed, in the Trail Smelter case, the United States and Canada had to ask the tribunal to specify regulations for the future control of emissions from the smelter and to assign monetary damages. No such precedents existed, and, curiously enough, few exist even today. No convention addressing international environmental protection goes beyond claiming that principles of liability should be developed.[20] No generally applicable norms or standards specify the terms of obligation, duty, accountability, or compensation; such issues are left to be decided in individual cases. As a general principle of international law, then, a liability regime establishes only that a state must warn others if actions within its own borders will harm its neighbors and, in the event of transborder environmental damage, it must compensate injured parties. It does not enumerate concrete, relevant regulations and requirements.

A more damaging drawback of liability regimes is that many environmental problems result from multiple sources or from a range of activities which in themselves are not hazardous but, due to widespread practices, have adverse cumulative effects. Global warming, for example, is a function of carbon dioxide, methane, chlorofluorocarbons, and other greenhouse gases which are released into the atmosphere from numerous sources. How would a liability regime begin to assign blame or sort out responsibility with regard to climate change? More generally, and to emphasize the point in a different way, how would liability regimes address problems that adversely affect global commons areas, vast regions of the globe, or the earth's carrying capacity? As the Earth's atmosphere or biological sustainability comes under increasing stress, and as the sources of environmental harm in general multiply and interact with each other, liability regimes, however sensible or successful in particular cases, will become increasingly less useful. They are, to put it differently, largely impractical in terms of bringing legal forms of governance to many types of global environmental degradation.[21]

Regulatory Regimes

In addition to liability regimes, the international system enlists regulatory institutions to govern activities that adversely affect the global environment. Regulatory regimes are essentially international agreements that set the terms of specific types of state actions. They aim to put into place goals for reducing environmental harm and establish specific practices toward the achievement of such goals. In contrast to liability regimes, they focus not only on instances of individual transboundary environmental harm but also on curbing environmental practices from a host of states that, taken cumulatively, degrade global commons areas or the overall infrastructure that supports life on earth. States assume little environmental responsibility unless they are legally assured that others will assume similar responsibility and be bound by those commitments. Regulatory regimes establish legally binding rules, procedures, and standards to enhance coordinated transboundary or global responsibility. They aim to harmonize state environmental policies with regard to problems that have no single identifiable source but which, nevertheless, are consequential. The bulk of international environmental efforts to date take the form of conventions based on a regulatory framework. The Montreal Protocol on Substances That Deplete the Ozone Layer, the Basel Convention on the Control of Transboundary Movements of Hazardous Wastes and Their Disposal, and the Convention on International Trade in Endangered Species (CITES) are all instances of such regimes.

International regulatory regimes have been crucial to protecting many aspects of the global environment. The Montreal Protocol and others have undeniably contributed to global environmental protection.[22] Moreover, according to some observers, regulatory agreements have worked to erode critical aspects of sovereignty and promote extraterritorial responsibility.[23] Arguably, however, notwithstanding such evaluation, regulatory regimes still privilege state rights over state responsibility because the prerogative that states enjoy to sign and adhere to agreements largely overshadows communal obligation. States have the right, for instance, to refuse even to become party to regulatory regimes. They can choose not to enter into international agreements and thus place themselves outside the jurisdiction of such regimes; they can become signatories but attach so many reservations that the regimes are hardly binding; or they can choose to abstain from individual decisions, as is permitted in the whaling regime and CITES. Additionally, even those states that agree to certain measures may refuse to relinquish sovereign rights when it comes to implementation. Once regulations are agreed upon in an international setting, the processes of interpretation, harmonization with national legislation, reportage, monitoring, and enforcement revert back to contracting states. There is no supranational organization to oversee compliance or enforce obligations; states resist any oversight mechanisms that might take the power to assess performance out of their hands. This, of course, greatly compromises implementation, and when violations occur there are very few mechanisms built into agreements for evaluating and adjudicating responsibility. Even the Montreal Protocol, recognized by many as the most successful environmental agreement on record,[24] fails to specify any nonnational body to review data and assign blame and penalty.

Liability and regulatory regimes are not, it should be clear, counterproductive or irrelevant for addressing transboundary or global environmental challenges. Both contribute tremendously to environmental protection. Nonetheless, each has certain limitations in its ability to negotiate a more equitable balance between state sovereignty and extraterritorial responsibility. These regimes have yet to find the appropriate balance between the two or even deliver substantially more weight to the side of obligation and accountability.

Sovereignty and Nonstate Mechanisms of Governance

The effort to construct liability or regulatory regimes is an attempt to generate global responsibility from *within* the state system. It represents the effort by states to establish rules, principles, norms, and so forth with which to bind themselves and garner greater global responsibility. The international system is constituted, however, not only by states but also by nonstate actors that attempt to project greater responsibility onto the state system. Moves from "outside," as it were, come predominantly from environmental NGOs. NGOs are organizations through which people voluntarily share certain interests or values, or express particular political aspirations, or a combination of both. What distinguishes them from states is that they are, generally speaking, what James Rosenau calls, "sovereignty-free actors."[25] This is especially the case with transnational NGOs. They are unconstrained by the host of concerns that revolve around territoriality; they focus and pursue aims free from the tasks of preserving and enhancing the welfare of a given, geographically situated population. Environmental NGOs are groups that work to conserve resources, reduce pollution, and protect the nonhuman world. To be sure, throughout much of the world the designation of being an "environmental" group may be misleading. Given the conjunction of human rights, poverty, development, public health issues, and environmental concerns, many NGOs advance a combination of causes that can loosely be associated with protecting the earth's ecosystem. The Harambee (*let's pull together*) in Kenya, for example, undertakes much rural development work with an aim toward sustainable development. Likewise, the Citizens Clearinghouse Network in the United States concentrates on protecting public health by focusing on the emission of toxic substances into the environment. Both can be considered environmental NGOs although they obviously have other concerns as well.

Environmental NGOs pursue a number of tasks, many of which aim to enhance global responsibility among states. At the most general level, NGOs work to promote environmental awareness among all people. Environmental problems arise cumulatively out of the actions of individuals, groups, corporate bodies, and states. Environmental NGOs target each of these levels of human experience in an attempt to orient practices in more environmentally sound directions. They thus work to shape economic structures, social mores, cultural understandings, and so forth with the idea that these are significant forms of governance that condition widespread behavior. They also recognize, however, that states play an important role in influencing human action throughout the world. While states do not command all forms of governance, they do possess a certain quality of power that enables them to influence human practices in an efficient manner. As Weber put it, states possess a monopoly over the legitimate use of violence within a given territory. As such, they can use law backed by force to reach into and affect the character of citizen action. For this reason, they represent a key node in the structure of power in the international system. NGOs recognize this leading role and, in addition to their society-oriented types of practices, work to influence state action.[26] In the context of this chapter, these actions can be understood as attempts to enhance global responsibility among states. They can be seen as efforts to diminish the self-regarding character of states formalized in the institution of state sovereignty and to intensify sensibilities and practices based on global obligation and accountability.

The main way that NGOs do this is through their roles in the formation, monitoring, verification, and modification of international environmental regimes. NGOs have an obvious role to play in drawing attention to environmental problems and thereby creating domestic and international pressure on states to create regimes that focus on transnational environmental issues. NGO efforts were central, for instance, to the creation of the Basel Convention, which outlaws the transportation of most toxic substances from OECD countries to non-OECD countries. At the transnational level, Greenpeace almost single-handedly investigated the so-called toxic waste trade and documented and publicized its grislier dimensions. At the regional level, organizations such as the Centre for Science and the Environment (CSE), based in New Delhi, helped alert citizens and governments in southern countries about the risks of importing toxic wastes and worked to generate official commitment to establish the Basel Convention. Both efforts were essential to the genesis of the Basel Convention.[27] Along these lines, it is important to point out that, in helping to establish certain environmental regimes, NGOs not only publicize issues but also set many of the actual terms of international accords. Greenpeace and CSE helped formulate language that eventually appeared in the Basel Convention, as do many NGOs with regard to a host of environmental accords. According to Jessica Mathews, NGOs set the original goal of a climate change convention and proposed most of its structure and content as it was negotiated at the Earth Summit.[28] NGOs were able to do so insofar as their representatives served on many governmental delegations, ensuring that their views about climate change and other issues were heard.[29]

In terms of monitoring and verification of environmental regimes, NGOs increasingly play both formal and informal roles in information gathering and reportage, both of which are currently inadequate.[30] In the case of CITES, for example, the International Union for the Conservation of Nature (IUCN) (with both government and nongovernment members) provides secretariat services on a UNEP contract.[31] Furthermore, in an unusual arrangement, IUCN officially delegates research, monitoring, and technical assistance functions to the World Wide Fund for Nature (known in the United States as World Wildlife Fund [WWF]).[32] This makes sense insofar as NGOs are often among the best-positioned actors to discover regime violations, publicize them to the press, and lodge formal complaints to treaty secretariats. Their activities often lead to a tightening of regime measures. For example, according to Peter Sand, since the inception of the European Union, more than half of the infringement proceedings relating to international environmental issues entered against member states were based on formal complaints from local and regional environmental NGOs.

Finally, with regard to modifying existing regimes, NGOs play a key role in tracking new scientific evidence as to the nature and intensity of environmental degradation, publicizing it, and working to upgrade regimes to reflect new environmental realities. Given the dynamic character of environmental change, international accords are almost always in need of revision. NGOs encourage such revision and have been responsible, in a few instances, for proposing the content of treaty upgrades. For example, after states established the Montreal Protocol on Substances That Deplete the Ozone Layer (brought about in large part because of NGO efforts in the United States and United Kingdom),[34] there was a need to revise national commitments due to new scientific evidence of an expanding ozone hole over Antarctica and new discursive frames for understanding the severity of the threat.[35] Friends of the Earth, Environmental Defense Fund, and other NGOs worked to persuade state officials to enhance the protocol. Their efforts were especially fruitful in establishing the 1990 London Upgrades to the protocol, which led eventually to the Copenhagen agreements that set the terms of a complete ban of ozone-depleting substances.[36] Another example of such efforts is the work by NGOs to scuttle a proposed minerals resources regime to be established in Antarctica as a part of a revision to the environmental protection dimension of the Antarctica Treaty System. Greenpeace, the Antarctica and Southern Ocean Coalition

(ASOC), IUCN, and other NGOs lobbied governments to reject the Convention on the Regulation of Antarctica Mineral Resources Activity (CRAMRA), a convention that would have opened up the continent to (controlled) mineral exploration. NGOs worked successfully to have CRAMRA not only rejected but replaced by the idea of a world park, codified by the protocol on Environmental Protection to the Antarctica Treaty.

The efforts by NGOs to create, implement, and enhance international environmental regimes represent attempts to build greater responsibility into international environmental affairs. Given that many steps in regime formation and implementation are inadequately performed by states, an increasingly active role is being claimed by NGOs. The sovereignty-free character of NGOs enables them to assume such a role and to undertake significant measures for seeking substantively strong regimes, engendering greater compliance, and upgrading accords when newly understood ecological realities and different conceptual frames require such action. NGOs, to be sure, do not solve the ecology/sovereignty problematique, especially insofar as many of their efforts work to inch the state system itself toward greater responsibility. Moreover, NGOs themselves can be subject to serious charges of arrogance and territoriality, as seen in the well-known conflicts among NGOs, including some serious ideological differences, North-South tensions, and questions of power, paternalism, and accountability. Nonetheless, their increasing role in all stages of regime activities makes them an important component within the international system, despite such internal divisions and problems, working to increase global responsibility among states and diminish the self-regarding orientation of states.

Reorientation of Sovereignty

State and NGO efforts to create and sustain environmental regimes are attempts to qualify state sovereignty. They work to bring the dual needs of rights and responsibilities within the state system more into balance by restricting state sovereign authority. They represent the institutionalization of what policymakers called for in both the Stockholm and Rio declarations. Further measures, however, need to be taken. Sovereignty still undermines responsibility, notwithstanding the many environmental regimes presently in place. Rather than solely focusing on qualifying sovereignty, then, some effort should be directed toward reorienting our understanding and practice of it. This needs to start with an appreciation of the conceptual straightjacket that conditions thinking about sovereignty and depiction of a promising trajectory for sovereignty's reorientation.

When most people entertain the idea of reformulating the notion of sovereignty, they almost immediately dismiss it because it entails, in their minds, some form of world government.[37] The notion of tinkering with a state's sovereign right to control events within its territory usually suggests a shift of authority away from individual units toward a supranational body. This perception greatly limits endeavors to fashion a form of governance to address global environmental problems. It suggests an either/or situation: Either states give up their rights to control domestic affairs or not. There is, seemingly, no middle ground.

One such innovation came in 1989, when seventeen heads of state signed the Declaration of the Hague. This almost universally neglected document sought to create a new, or newly strengthened, body within the UN that would be able to make decisions concerning environmental issues even in the absence of unanimous agreement. The signatories, which included Brazil, France, India, Japan, and the former West Germany, understood that the requirement of unanimity was a major stumbling block for designing and implementing regulations to protect the earth's ecosystem. It allowed certain states to remain outside the jurisdiction of international regimes and thus undermined efforts to protect the global environment. The signatories wanted to go beyond the idea that states can be bound to international agreements only by their own consent, and called for binding regimes in the absence of unanimity. The UN body to oversee such regimes would have much power. It would be able to impose penalties for violations of existing regimes and mandate that international environmental disputes be brought to the World Court for arbitration.[38] Referring to the document, Prime Minister Gro Brundtland of Norway commented, "The principles we endorsed were radical, but any approach which is less ambitious would not serve us."[39] The innovative character of the document turns on its attempt to think beyond the narrow strictures of state sovereignty without suggesting a form of global government.

As the world moves into the post-Rio era, there is a need to look toward initiatives like the Hague Declaration for creative ways to reconceptualize and reorient sovereignty. While it is beyond the scope of this paper to formulate such a reconceptualization, it is worth pondering what a reoriented sense of state sovereignty might mean. First, there is the idea represented by the Hague Declaration of creating a supranational organization having authority over individual states with regard to environmental matters but not possessing the capacity for coercive control advocated by most proposals for world government. This goes against the notion, enshrined in all international law, that states can only be bound to international agreements by their own consent. Empowering an institutional body to enforce certain environmental mandates — even if states do not consent to them — is a way of chipping away at sovereignty within a specific issue area, without abandoning the principle itself. Such an institution would be, in the words of Richard Falk, a "centralized guidance system"[40] — a mechanism to oversee and exercise authority, but one that is denied the coercive control to govern actual domestic affairs.

Reformulation does not, however, only have to refer to a supranational authority. Additionally, one can imagine folding internationalizing elements into the reporting, monitoring, and compliance dimensions of environmental practices, a process in which internationally oriented NGOs could become closely involved. For example, UNCED established the practice that national governments are responsible for submitting reports on their progress in implementing measures mandated by UNCED. This allows national governments to retain oversight for environmental actions themselves. However, as many have pointed out in the context of other international issues, this approach resembles asking the wolf to guard the chicken coop. If governments themselves recount their actions, there is always the possibility of inaccurate reporting for political purposes. One way to guard against this would be to include NGOs from the home country — which are often independent of the government and therefore likely to be more critical — to participate in reportage. Governments could be required to work with legitimate nongovernmental elements within their own countries in preparing and submitting country reports.⁴¹ A number of countries already do this. Further inclusion of NGOs in the process could ensure that governments submit authentic reports uninfluenced by attempts to win political advantage. Of course, there are many states that deprive their citizens of the opportunity for forming genuinely independent NGOs. Indeed, authoritarian regimes are some of the worst environmental offenders and have no independent "third sector" to play such a role. In these countries, the inclusion of domestic NGOs in environmental assessment and reporting is, to be sure, greatly hampered. To ameliorate this situation and to enhance reporting more generally throughout the world, transnational NGOs—organizations with offices in more than one country and thus less likely to represent the interests of a given territory—should be included in the process. Organizations such as WWF and IUCN are well suited for such work and would add another element of impartiality.⁴² Again, such inclusion would not be easy in authoritarian regimes—especially if one imagines a completely cooperative relationship. Nonetheless, many transnational NGOs already monitor environmental affairs within authoritarian countries. What needs to be done is to allow the data these organizations collect to be formally accepted by international environmental organizations. All such efforts aim to stretch the process of environmental diplomacy, as it relates to submitting data to international organizations, across state/society lines and across national borders, thereby deterritorializing environmental politics.

Another way to refashion the practices of sovereignty could entail on-site inspections by mutually friendly nations. At present, states retain the role of monitoring their own behavior. This is a problem insofar as it includes no outside parties to ensure accountability. While other countries and UN bodies can monitor large-scale environmental harm through satellite observation, the limits of such technology call for periodic, complementary, on-site inspections. In the past, with regard to certain arms control agreements, states have objected to on-site assessments because of possible leakage of national security secrets. This is partially the reason for rejecting any intrusive monitoring with regard to environmental issues. To work against this, inspections could be conducted by allies or friendly states. For example, an inspection team from the United Kingdom, France, and Israel might observe U.S. actions, or a team from Brazil, India, and Indonesia could monitor Malaysia's actions. Such a process might remove some of the hesitations about inspections in general. An "amity inspection regime" or some variation of it would limit sovereignty with regard to environmental issues without completing eroding it. Moreover, making domestic actions more transparent to international observation would help enhance global responsibility.

Again, the problem of closed societies emerges. Would sovereignty genuinely be reoriented if Burma monitors Vietnam or China, for instance? How do amity regimes work in such settings? To be sure, in these cases, one would instantiate only a small measure of transparency. Nonetheless, it would still crack the container circumscribed legally by sovereignty and begin a process of dialogue and interaction based on others helping to determine domestic environmental conditions.

A more ambitious program would be a system, ideally global in scope, of labels on certain products describing their region of origin and the environmental costs associated with their production, transport, packaging, and use. Such a program would put a spotlight on states' "shadow ecologies"—areas of the Earth outside a state's boundaries from which it draws resources and exports wastes.[43] The doctrine of sovereignty helps perpetuate the illusion that the ecological costs of daily life affect only one's national territory. An environmental labeling system would begin to challenge those understandings and make people more aware of the interdependence of countries in resource use and environmental degradation. Such a program would be similar to that of food labeling, which now serves as an educational device to improve personal nutrition. Today, many countries have laws demanding that food products list ingredients and nutritional information on packaging. While the initial intent of such laws had to do with safety provisions associated with adulterated, misbranded foods, they have since become mechanisms for health and nutrition standards and education.[44]

A similar process could take place with regard to environmental costs. Labeling could begin as an attempt to raise awareness about the transboundary dimension of resource use and the fragility of the earth's bounty, and evolve into a system by which people take action to preserve the earth's ecosystem. The most obvious direction would be toward what economists call "full-cost pricing"—a system that seeks to capture the actual costs of commodity production and use. Such a system would incorporate the amount of detrimental environmental damage associated with production, transportation packaging, use, and disposal into the price of products.[45] The internationalization of such a system would work against the parcelling of the Earth's environment into separate, seemingly autonomous jurisdictions and could help to inculcate a sense of global responsibility as people understood and took actions to respect the geographical range of their resource use.

These and other mechanisms would open up the processes of environmental politics to international scrutiny and thus increase incentives for globally responsible action. Responsibility thrives in a context of accountability, and the above ideas aim foremost to increase oversight or at least outside review. While they do not create a supranational body that would actually supervise domestic environmental policy, they work toward dissolving the shell of sovereignty that allows unscrutinized behavior.

While these proposals focus mainly on issues of shared resources or protection of global commons areas, some effort is also needed in changing domestic practices that have no direct effect on other countries but which nonetheless degrade the overall carrying capacity of the Earth. In this regard, an attempt is needed to dislodge the stubborn and detrimental effects of a doctrine of sovereignty understood as sanctuary from global responsibility.[46] This would not entail dismantling states' sovereign rights to oversee and administer their territories, but would rather emphasize, in language, deeds, and institutional support, the responsibility that should accompany sovereign rights. Put differently, it could entail freeing diplomacy from the shadow of the UN Charter, which proclaims that states ultimately retain control over their own domains. Such liberation would precipitate a new set of understandings with regard to responsibility. With such a shift, Principles 21 and 2 might read very differently. Instead of reaffirming states' sovereign rights over their territories, the principle could state:

Contrary to the tradition of interstate practices, states do not have the sovereign right to exploit their own resources pursuant to their own environmental and developmental policies. Rather, they must reflect upon, coordinate, and harmonize with other states those activities that adversely effect the environment. This will ensure that they genuinely assume responsibility for the transboundary and global effects of their activities.

To be sure, such a principle would not revolutionize the nature of sovereignty, nor would it fully usher in an era of state responsibility for the global ecosystem. It would, however, begin to bring international understandings and practices in line with the nature of contemporary ecological realities.

Conclusion

The United Nations Conference on Environment and Development held in Rio de Janeiro in 1992 was an important moment in modern environmental history. Like its counterpart twenty years earlier, it provided a focal point for human enterprise in relation to international environmental protection. As such, however, it has only raised the pitch of environmentalism without undertaking the huge task of fundamentally redirecting human practices. According to Lester Brown, the task of global environmental restoration and protection is on the order of a revolution. "If this Environmental Revolution succeeds," he writes, "it will rank with the Agricultural and Industrial Revolutions as one of the great economic and social transformations in human history."[47] One reason such a change would be enormous is that it would, of necessity, entail a fundamental reformulation of the concept of national sovereignty and a burgeoning of a sense of global responsibility among all states.

I have aimed to analyze the nature of that challenge by focusing on the relationship between state rights and responsibilities as they turn on contemporary formulations of national sovereignty. I have reviewed the way states themselves and NGOs have tried to relax sovereign authority in the interest of transnational environmental protection. Furthermore, I have explored the promise of certain policies that would tilt the balance between rights and responsibilities more in the latter's direction. While such proposals may sound unrealistic or unwieldy, there is an urgent need to think beyond the formulas encoded in existing international law and expand the perimeters of conventional practices. In the post-Rio era, when we have finally come to understand the broad outlines of global environmental degradation, much depends upon our doing so.

Notes

1. Aldo Leopold, *A Sand Country Almanac and Sketches Here and There* (New York: Oxford University Press 1989): 35.

2. See Thomas J. Biersteker and Cynthia Weber, eds., *State Sovereignty as a Social Construct* (New York: Cambridge University Press, 1996).

3. Ken Conca, "Rethinking the Ecology-Sovereignty Debate," *Millennium* 23, no. 3 (Autumn 1994).

4. The distinction between global commons, shared resources, and transboundary externalities is taken from Oran Young, *International Governance: Protecting the Environment in a Stateless Society* (Ithaca, N.Y.: Cornell University Press, 1994): 20–26.

5. For discussions of the relationship between sovereignty and international environmental issues pre–1972, see Albert Utton, "International Water Quality Law," in Ludwik Teclaff and Albert Utton, eds., *International Environmental Law* (New York: Praeger, 1974. See generally, Barry Carter and Philip Trimble, *International Law* (Boston: Little, Brown, 1991): 1145–1210.

6. 21 Op. Attorney General 281 (1885). Quoted in Lynton Caldwell, "Concepts in Development of International Environmental Policies," in Teclaff and Utton *International Environmental Law*, 21.

7. Carter and Trimble, *International Law*, 1149.

8. Thomas Naff and Ruth Matson, *Water in the Middle East: Conflict or Cooperation?* (Boulder, Colo.: Westview, 1984), 165–66.

9. The cases usually cited as bringing about this evolution include: 1909 Boundary Waters Treaty between the United States and Canada, 1941 Trail Smelter Arbitration, 1949 Corfu Channel, and 1957 Lake Lanoux Arbitration.

10. For a discussion of Principle 21 and the issue of sovereignty, see Louis Sohn, "The Stockholm Declaration on the Human Environment," *Harvard International Law Journal* 14, no. 3 (Summer 1973): 485–93.

11. United Nations, *Report of the United Nations Conference on the Human Environment*, UN Doc. A/Conf.48/14/Rev.1 1973, 5.

12. Allen Springer, "United States Environmental Policy and International Law: Stockholm Principle 21 Revisited," in John Carroll, ed., *International Environmental Diplomacy: The Management and Resolution of Transfrontier Environmental Problems* (Cambridge: Cambridge University Press, 1988): 50.

13. See, for example, Recommendations 3, 37, 48, 51, 70, and 92 in United Nations, *Report of the United Nations Conference on the Human Environment*, UN Doc. A/Conf.48/14/Rev.1 1973.

14. The only difference, although an important one for other reasons, is that the Rio Declaration emphasizes development needs.

States have, in accordance with the Charter of the United Nations and the principles
of international law, the sovereign right to exploit their own resources pursuant to
their own environmental *and development* policies, and the responsibility to ensure
that activities within their jurisdiction of control do not cause damage to the
environment of other States or of areas beyond the limits of national jurisdiction.

15. Fred Pearce, "How Green Was Our Summit?" *New Scientist* 134, no. 1827: 13.

16. Wolfgang Sachs, "Global Ecology and the Shadow of 'Development,'" in Wolfgang Sachs, ed., *Global Ecology: A New Arena of Political Conflict* (London: Zed Books, 1993): 3. See also, Christopher D. Stone, *The Gnat is Older than Man: Global Environment and Human Agenda* (Princeton, N.J.: Princeton University Press, 1993): 108. According to some, downgrading from a treaty to a set of nonbinding principles worked to prevent Northern countries from dictating unfavorable and perhaps even antienvironmental terms to Southern states. This point was made to me in personal correspondence with Anil Agarwal, May 1996, and finds justification in Anil Agarwal and Sunita Narain, *Towards a Green World* (New Delhi: Centre for Science and Environment 1992): 71–84.

17. Quoted in Fred Pearce, "Earth at the Mercy of National Interests," *New Scientist* 134, no. 1826: 4.

18. Stone, *The Gnat is Older than Man*, 58.

19. Carter and Trimble, *International Law*, 1150–59.

20. Patricia Birnie, "International Environmental Law: Its Adequacy for Present and Future Needs," in Andrew Hurrell and Benedict Kingsbury, eds., *The International Politics of the Environment* (Oxford: Clarendon Press, 1992): 67.

21. Stone, *The Gnat is Older than Man*, 62–63.

22. See generally, Peter Haas, Robert Keohane, and Marc Levy, eds., *Institutions for the Earth: Sources of Effective International Environmental Protection* (Cambridge, Mass.: MIT Press, 1993); Young, *International Governance*; Peter M. Haas, *Saving the Mediterranean: The Politics of International Environmental Cooperation* (New York: Columbia University Press, 1990); Ronald B. Mitchell, *International Oil Pollution at Sea* (Cambridge, Mass.: MIT Press, 1994); and Lynton Caldwell, *International Environmental Policy*, 2d ed. (Durham, N.C.: Duke University Press, 1990).

23. See, e.g., Peter Haas with Jan Sundgren, "Evolving International Environmental Law: Changing Practices of National Sovereignty," in Nazli Choucri, ed., *Global Accord: Environmental Challenges and International Responses* (Cambridge, Mass.: MIT Press, 1993).

24. Richard Benedick, *Ozone Diplomacy* (Cambridge, Mass.: Harvard University Press, 1991): 1.

25. James Rosenau, *Turbulence in World Politics: A Theory of Change and Continuity* (Princeton, N.J.: Princeton University Press, 1990): 37.

26. For an extensive discussion of the society-oriented strategies of environmental NGOs, see Paul Wapner, *Environmental Activism and World Civic Politics* (Albany, State University of New York Press, 1996).

27. See, generally, Willy Kempel, "Transboundary Movements of Hazardous Wastes," in Gunnar Sjostedt, ed., *International Environmental Negotiation* (Newbury Park: Sage, 1993; Greenpeace, *Toxic Trade Update* 6 no. 2 (April 1993); and Brian Wynne, "The Toxic Waste Trade: International Regulatory Issues and Options," *Third World Quarterly* 11, no. 3 (July 1989): 120.

28. Jessica T. Mathews, "Power Shift," *Foreign Affairs* 76, no. 1 (January-February 1997): 55.

29. Ibid.

30. General Accounting Office, U.S. Congress, *International Environment: International Agreements Are Not Well Monitored*, GAO/RCED-9-43, 1992.

31. J. H. Ausubel and D. G. Victor, "Verification of International Environmental Agreements," *Annual Review of Energy and Environment*, no. 17 (1992): 13.

32. Oran Young, *International Cooperation: Building Regimes for Natural Resources and the Environment* (Ithaca, N.Y.: Cornell University Press, 1989): 26. See also, Caldwell, *International Environmental Policy.*

33. Peter Sand, *Lessons Learned in Global Environmental Governance* (Washington, D.C.: World Resources Institute, 1990).

34. Wapner, *Environmental Activism and World Civic Politics*, 127–28, 132.

35. Karen Litfin, *Ozone Discourses: Science and Politics in Global Environmental Cooperation* (New York: Columbia University Press, 1994).

36. Barbara Bramble and Gareth Porter, "Nongovernmental Organizations and the Making of U.S. International Environmental Policy," in Hurrell and Kingsbury, *The International Politics of the Environment*, 341.

37. For example, Keohane, Haas, and Levy write:

> The skeptics are sure right to warn that the planet's ecosystem is in danger and that its protection will require modifications in traditional interpretations of state sovereignty. Yet, world government is not around the corner: Organized international responses to shared environmental problems will occur through cooperation among states, not through the imposition of government over them.

Keohane, Haas, and Levy, "The Effectiveness of International Environmental Regimes," 4. See also, Young, *International Cooperation*, 4. Introduction to *Institutions for the Earth* (Cambridge, Mass.: MIT Press, 1993).

38. See Hilary French, "After the Earth Summit: The Future of Environmental Governance," *Worldwatch Paper 107* (March 1992); and Sand, *Lessons Learned in Global Environmental Governance.*

39. The Centre for Our Common Future, "Background Information on the Hague Declaration," press release, Geneva, undated, quoted in French, "After the Earth Summit," 35.

40. Richard Falk, *A Study of Future Worlds* (New York: Free Press): 156. For a more recent formulation of the same concept see, Richard Falk, *Humane Governance: Toward a New Global Politics* (University Park: Pennsylvania State University Press, 1996).

41. In chapter 38 of Agenda 21, governments are encouraged, but not mandated, to work with NGOs in many capacities. See chapter 38.5–38.8, United Nations, *Report of the United Nations Conference on Environment and Development.*

42. WWF and a number of other groups are already involved in such processes. WWF and Earth Island Institute, for example, play a central role in monitoring CITES. See Wapner, *Environmental Activism and World Civic Politics.*

43. Jim MacNeill, Pieter Winsemius, and Taizo Yakushiji, *Beyond Interdependence: The Meshing of the World's Economy and the Earth's Ecology* (New York: Oxford University Press, 1991).

44. Compare, for example, the U.S. Federal Food and Cosmetic Act of 1938 with the Nutrition Labeling and Education Act of 1990).

45. See, for example, Lester Brown, "Launching the Environmental Revolution," in Lester Brown et. al., *State of the World 1992* (New York: W.W. Norton, 1992); Stephen Schmidheiny, *Changing Course: A Global Business Perspective on Development and the Environment* (Cambridge, Mass.: MIT Press, 1992): 14–33. For a critical discussion of full-cost pricing as an answer to environmental degradation, see Herman Daly, *Steady-State Economics* (Washington, D.C.: Island Press, 1991): 43ff.

46. See Shridath Ramphal, *Our Country, the Planet: Forging a Partnership for Survival* (Washington, D.C.: Island Press, 1992): 23.

47. Brown, "Launching the Environmental Revolution," 174.

12

Global Village Sovereignty: Intergenerational Sovereign Publics, Federal-Republican Earth Constitutions, and Planetary Identities

Daniel Deudney

Introduction[1]

Over the last several decades humanity has been confronted with a complex and growing crisis in its relationship with the natural environment. Environmentally abusive practices, compounded by the sheer weight of human numbers and the power of industrial technology, have increasingly degraded a wide range of vital and interconnected resources that humans had been able to take for granted throughout history. As bad as things have become, they promise to get much worse in the first decades of the next century: Human population is likely to double again, and economic output is likely to increase threefold to fivefold. These developments presage either the collapse of industrial civilization or a far-reaching "green" transformation of all aspects of human life.

The Global Village and World Political Theory

In the face of these developments, the quest for a new political paradigm to conceptualize solutions and frame agendas has immense practical importance. But what should this paradigm be? Perhaps the simplest and most resonant metaphor for the emergent human situation is Marshall McLuhan's "global village." This has become something of a cliché, but is actually a conceptually explosive oxymoron. From the standpoint of the Western tradition of political science and practice centered on the polis, the state, and the nation-state, the prospect of village-like proximity and interconnectedness occurring on a planetary scale demands far-reaching theoretical and practical innovations.

To conceptualize the forms of political association appropriate to a global village it is necessary to move beyond social science and return to the fundamental questions of political theory recast as world political theory. Social scientists seek to explain what has happened, but can offer only partial and indirect assistance in responding to the historically unprecedented situation of the global village. Grappling with the pressing practical problems emergent in the rapidly globalizing world requires a return to the type of creative and architectonic theorizing that characterized the project of political science and theory prior to the ascendancy of behavioral social science. It is also necessary to move beyond the antiquarian and marginalist political theory so prevalent in academic circles and rethink, reconfigure, and reemploy the basic conceptual components of political theory and practice.

While boldness is urgently needed, the guiding motto of the world political theorist must be "reuse, repair, and recycle," rather than to begin from scratch. If the task is to rebuild the political ship while at sea, the existing timbers will have to be reassembled rather than completely replaced. Or, to shift analogies, contemporary political garments may be full of holes and grossly ill-fitting, but they contain a wealth of strong threads available for reweaving and reassembling. All practical political theory works with the materials at hand rather than the ideal or the completely new.

It is also particularly important to build wherever possible from the conceptual materials present in the most successful and most hegemonic political traditions and vocabularies. With the exhaustion and collapse of systemic alternatives and competitors to Western liberalism in the last quarter of the twentieth century, this means rethinking and recombining the many disparate pieces of the commonwealth tradition of liberal, federal, republican constitutionalism in ways responsive to the practical problems of political association in the global village.

From Westphalian to Terrapolitan Sovereignty

The problem of sovereignty looms particularly large in the project of world political theory. In contemporary theory and practice *sovereignty* refers first and foremost to the Westphalian system of mutually recognized autonomous states with sole and final authority over a delimited territorial space. The Westphalian system of state sovereignty is highly contested in both practice and theory. For many, the state-centered sovereignty system is valued as a bulwark against domination or influence by outside cultural, economic, or political forces which globalization is intensifying.[2] Others hold that the sovereign state system is a major impediment to international and collective problem solving, and is incompatible with the realization of a range of important values, particularly security from violence, human rights, equitable economic development, and ecological sustainability.[3] These divergent assessments have been paralleled in the analyses of social scientists who are in fundamental disagreement about whether the state-centric sovereignty system is waxing, waning, or changing in world politics.[4]

The emergence of major environmental problems and efforts to address them call into question in a very radical way the modern Westphalian sovereignty system because the establishment of environmentally sound ways of life entails the establishment of new configurations of legitimate authority. Systems of authority are particular configurations of restraints and empowerments on individuals and groups. The Westphalian sovereignty system empowers states and restrains nonstate actors. Environmentalist agendas and discourses are hardly univocal, but a fundamental claim of contemporary ecological environmentalism is that natural limits have been transgressed, and that human actors at all levels and scales of world politics must either exercise self-restraint, enter into agreements of mutual self-restraint, or be subject to external restraint.[5]

Diminishing the privileged role of state claims to authority and autonomy thus occupies a pivotal role in the environmentalist effort to establish new forms of authority. Most of this effort centers upon ways to add "responsibilities" to the "rights" of sovereign statehood, to extend sovereign rights to groups deemed more ecologically responsive, or to induce or facilitate sovereign states either to exercise their powers according to norms or rules of constraint, or to divide, share, or pool their authoritative powers in collective restraint systems.[6]

Most of the focus of environmentalist analysts and practitioners has been deconstructive (challenging or overthrowing), or reformist (modifying) the state-centric sovereignty system,[7] rather than constructive of a fundamental alternative. For most environmentalists, sovereignty is something to be overcome, modified, or captured, rather than something that offers positive insight and guidance for building a sustainable society.[8]

This mainly negative orientation of environmentalists toward sovereignty is somewhat anomalous. Virtually every other major conceptual component of the Western tradition of political thought (democracy, individual and human rights, market capitalism, socialism, and communitarian anarchism) has been analyzed as a source of environmental problems, and reconceptualized and reclaimed as having a potentially positive role to play in environmentally sustainable governance.[9] The underdevelopment of a "green sovereignty" is partially the result of the fact that sovereignty is taken to be inherently synonymous with the autonomy and supremacy of the state, rather than recognized as a more protean concept useful for capturing the generative logic of legitimacy, authority, and identity in political orders more generally.[10]

There are, however, important strands of environmentalist practice, discourse, and theory that offer raw material for the conceptualization of authority and identity patterns that are consistent from the "ground up" with concerns of ecological responsibility. Over the last several decades, political theorists, ethicists, and theologians have reexamined in elaborate, sophisticated, and far-reaching ways the relationship between political, ethical, and spiritual systems and traditions, on the one hand, and environmental decay and restoration, on the other. Three of the largest clusters of creative ferment have centered upon "green democracy," "ethical extension," and what might be termed "nature theology." On the topic of democracy and the environment, theorists have rethought the roles of participation, representation, decentralization, and accountability in the light of sustainability imperatives. A hallmark of these efforts has been an emphasis upon revitalizing the "grass roots" dimensions of self-governing political associations. By far the largest and most conceptually developed body of work on ethical extension has been in terms of the Western tradition of rights and utility-based ethics. Specific arguments have been advanced to extend rights to nature as a whole, particular parts of nature, individual biological organisms, ecosystem integrity, and intergenerational groups.[11] A third large cluster of thinking addresses issues of theology, cosmology, and the sacred, and their relation to ecology and sustainability.[12] A common feature of these efforts is that submerged or marginalized components of the dominant traditions are refurbished and employed in new ways.

Notably absent from these political, ethical, and theological debates has been any positive reconstruction of sovereignty. The implications of green democracy, ethical extension, and nature theology for international politics and large-scale political associations remain to be drawn. To discern the essential political implications of green democracy, ethical extension, and nature theology, it is necessary to examine their possible role in constituting sustainable patterns of legitimate political authority. The missing link between green democracy, ethical extension, and nature theology, and world politics is a reconfigured "green sovereignty."

This chapter sketches the main elements of a theory of sovereignty appropriate to the governance imperatives of the emergent global village. It is obvious that such a conception of the basis of political association must move beyond the pure particularity of the Westphalian state and nation. But it is also necessary to steer clear of the universal, homogenous, and unmediated conceptions of cosmopolitan political association. While the global village is as extensive as a global cosmopolis, it is as immediate and intensive as a village, and must accommodate more complexity and diversity than ever before has been encompassed in single political association. In short, a conception of political authority and community appropriate to the novel configuration of the global village must be *terrapolitan* rather than either Westphalian or cosmopolitan. By this I mean that the central basis of political association in the global village must be the Earth (terra) and its requirements.

This chapter sketches the rudiments of a terrapolitan conception of sovereignty, legitimate political authority, and communal identity. The heart of the argument is simple and has two main parts. First, sovereignty situated in an intergenerational public provides the basis for a federal-republican Earth constitution. Second, Earth nationality and Gaian Earth religion provides the basis for community and identity necessary to instantiate and maintain this sovereign and the legitimate authorities consistent with it.

The formulation of this argument entails a mixture of conceptual, deductive, and empirical elements. It aims not to provide a blueprint of an ideal terrapolitan arrangement, but rather to reformulate neglected and suppressed components of the commonwealth tradition of Western political thought in ways that mesh with powerful, if disparate, currents in actual environmental practice and in ways that address fundamental practical problems in achieving ecological sustainability.

The argument proceeds in three main steps. The first section, employing the traditional political theoretical distinction between sovereignty and authority, outlines the general relationship between particular sovereigns and particular forms of political authority and communal identity. Then the argument focuses on one particular form of sovereignty — popular sovereignty — and its relationship to authoritative governance that is republican and constitutional, and its problematic relationship to communal identity. The second section employs John Dewey's concept of the public to sketch the nature of an intergenerational sovereign public, and then argues its necessary relationship to a republican Earth constitution. The third section argues that Earth nationalism and Gaian Earth religion together constitute the logical and appropriate communal identities necessary to instantiate and manifest an intergenerational sovereign public.

Sovereignty, Legitimate Authority, and Communal Identity

In thinking and talking about sovereignty, much confusion has arisen from the failure to distinguish consistently between sovereignty, authority, and particular patterns of authority. *Sovereignty*, in its original and basic meaning, is simply the ultimate and undivided source of all legitimate authority in a polity, while *authority* refers to the actual exercise of legitimate power. Following Bodin and Hobbes, William Blackstone formulated the classic definition of sovereignty: "There is and must be in all [forms of government] a supreme, irresistible, absolute, uncontrolled authority, in which the *jura summi imperii*, or the rights of sovereignty, reside."[13] In terms of these distinctions, the external practice of mutual recognition in the Westphalian anarchical society — the state "sovereignty" of international public law — is a particular form of authority.

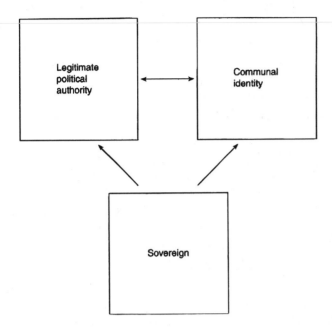

Figure 12.1
The sovereign and patterns of legitimate authority and communal identity

The distinction between sovereignty and authority is necessary in order to frame the two decisive questions at the heart of debates about sovereignty: (1) Who—or what—is sovereign? and (2) Which patterns of legitimate authority and communal identity are consistent with a particular sovereign? (See figure 12.1.)

The first question concerns the location of sovereignty, and it has been answered in a variety of very different ways. Over the last five hundred years in the West, strong claims have been asserted for the sovereignty of God, kings, states, nations, consumers, and the people as a whole.

Different sovereigns are the sources of different patterns of legitimate authority and communal identity. Once the location of sovereignty has been asserted or posited, it becomes possible and necessary to examine the relationship between particular sovereigns and particular patterns of legitimate authority and communal identity. Assertions of sovereignty amount in practice to claims that particular patterns of authority are legitimate, and patterns of identity are authentic. Each sovereign entails a particular configuration of such authorities and identities.

To ascertain which patterns of authority and communal identity manifest and instantiate a particular sovereign, one must examine the fundamental interests of the sovereign body. All possible sovereigns share two fundamental problems: (1) securing the sovereign body from destruction, and (2) avoiding usurpation—the appropriation of sovereignty by some other entity. Thus arise distinctive combinations of sovereignty and patterns of legitimate authority and communal identity. Beyond this generality, the specific features of the patterns of authority and communal identity consistent with a particular sovereign can only be determined by examining the attributes of the sovereign and the particular significant threats to its security. For example, assertions that God or some particular religiously privileged agent or group is sovereign generate a hierarchical structure of legitimate authority (*hier = priest*), and a communal identity embodied in and reproduced by a "church."

Popular Sovereignty and Federal-Republican Constitutions

Of all the principal sovereignties that have been asserted in the course of Western political history, popular sovereignty is closest to the intergenerational sovereign public emergent in planetary environmental practice and discourse. Therefore, to lay the foundation for the extension of sovereignty to an even broader group, it is useful to review briefly the central arguments in the republican political tradition about the particular forms of legitimate authority and communal identity consonant with popular sovereignty.

The claim that the people as a whole are sovereign was first asserted with significant political consequence in early modern Europe.[14] The Bostonian revolutionary writer James Otis framed the assertion of popular sovereignty with particular vigor:

An original, supreme, absolute, and uncontrollable earthly power must exist in and preside over every society, from whose final decisions there can be no appeal but directly to Heaven. It is therefore, originally and ultimately in the people; . . . and [they] never did in fact freely, nor can they rightfully, make an absolute renunciation of this divine right. It is ever in the nature of the thing given in trust, and on a condition the performance of which no mortal can dispense with, namely, that the person or persons on whom the sovereignty is conferred by the people, shall incessantly consult their good.[15]

Popular sovereignty formed the basis for the complex of legitimate authorities of "liberal" — federal, republican, democratic, and constitutional — political orders. It is important to emphasize that popular sovereignty is equivalent to democracy in the sense of majority rule. For a popular sovereign to manifest itself as a democracy is problematic because this entails a division of the people into at least two parts (the ruling majority and the ruled minority), and sovereignty is by definition indivisible. If a democratic majority has all the authority in a political order, a hierarchy in the form of a "tyranny of the majority" exists between the majority and minorities. Only in the demanding circumstances of enforced homogeneity sketched by Rousseau in the *Social Contract* is popular sovereignty consistent with democracy.

A central claim in the republican political tradition is that sovereignty situated in the people as a whole must be manifested in political authorities structured so as to restrain hierarchical political authorities. This is particularly true if the people are so numerous and dispersed as to require extensive delegations of power. Popular sovereigns give rise to "republics" in the sense of governments and identities of restraint.[16] Unlike restraint based on hierarchical domination, republican political orders are complexes of mutual power restraint. At the heart of the system of structural power constraints that arise from a popular sovereign are written constitutions that can only be altered by supermajorities acting over extended periods of time, and judicial systems that have institutionalized vetoes over legislative and executive actions.[17]

The Problem of Community, Identity, and Virtue in Republican Orders

The expression of a particular sovereign in authoritative and legitimate governance structures is an incomplete basis for a political order because sovereigns also require manifestation and instantiation in appropriate forms of communal identity. Three essential points about the types of community and identity consistent with an extended popular sovereign and republican constitutions are well known and can be summarized briefly.

First, the primary community and identity feature associated with popular sovereigns and republican constitutions is civic, meaning that citizens have identities divided along a spectrum of private, semiprivate, and civic (or public).[18] In the private and semiprivate realms, identity differences of great magnitude exist but are buffered or compartmentalized from direct political roles. Civic identity and community are much more universalistic and attenuated in their claims upon individuals. In the modern republican model, patterns of communal identifications and practices are mixed and layered, thus allowing for the partial expression of intense and particularistic communal identities within the private and semiprivate spheres. While lacking the purity and total quality of communal identities in smaller and less diverse polities, this pattern of communal identity has one overriding virtue from the standpoint of a popularly situated sovereignty: the ability to encompass very large numbers of individuals that do not otherwise share common identities.

This pattern of liberal and multiethnic communal identity has, however, two fundamental and chronic problems: maintaining a sufficiently strong sense of common community — fraternity — and inculcating self-restraint — "virtue" — in individuals and groups. Political orders based upon popular sovereigns have not been highly successful in generating their own sources of fraternity and virtue, but have come to depend upon other sources, most importantly national patriotism and civic religion.

In the absence of some sense of fraternal[19] identity, republican political orders are subject to the problems of extreme factionalism.[20] National patriotism has served to provide republican political orders with a sense of communal solidarity that they are not able to generate on their own.[21] Individual and group discipline or self-restraint (virtue) is the second vital, but precarious, component of republican identity. Decentralized political structure and absence of a coercive hierarchy are only possible when the people discipline themselves. Without popular self-discipline, conflicts would be too sharp and passions too extreme to be mediated by the republican constitutional governmental structures. Republican virtue also meant putting the interests of the public over private ones. Republican political theorists have been thus intently concerned with establishing and sustaining virtue against its antithesis, corruption.[22]

The long record of illiberal and hierarchical political orders demonstrates the ease with which humans may be socialized or disciplined into obedience. But it is more difficult for individuals to gain the psychological practices (or in the language of moral philosophy, "character") of self-discipline that are not reliant upon deference to traditional or hierarchical authorities. Despite the emphasis of republican political theorists on the need for civic education, particularly militias, the key sources of virtue in large, modern, republican constitutional political orders have also been largely outside the programs and practices of republicanism. If Tocqueville and his many followers are correct, republican political orders require particular forms of religion, initially in Protestant Christianity, but subsequently in "reformed" versions of Catholicism and Judaism.

The dependence of popular sovereigns and their republican constitutional structures on certain forms of nationalism and religion poses an even greater problem for an intergenerational public sovereign and a republican Earth constitution, but before these problems and possible solutions to them can be fully understood, it is necessary to examine the notion of an intergenerational public sovereign and its relationship to a republican Earth constitution.

Intergenerational Sovereign Publics and Republican Earth Constitutions

The tools are at hand with which to conceptualize the relationship between a popular sovereign and the patterns of legitimate authority and communal identity consistent with it, but what possible relationship can there be between this federal-republican approach to political association and the environmental problematique arising to define and besiege the global village? The key to this link in the argument is John Dewey's conception of the public. The greatest American political theorist in the first half of the twentieth century, Dewey sought to reconceptualize the nature of democracy in face of the changing material environment produced by the Industrial Revolution and in doing so introduced formulations directly relevant to the emergent global village.

Intergenerational Sovereign Publics

Like most of the important terms of political discourse and theory, *public* has assumed a range of competing meanings and usages. Most commonly *public* refers to the nonprivate realm in liberal political orders, or to the people as a whole. More useful in conceptualizing a green sovereign is John Dewey's spare and distinctive definition: "The public consists of all those who are affected by the indirect consequences of transactions to such an extent that it is deemed necessary to have those consequences systematically cared for."[23] The public is a human consequence group, which Dewey sharply distinguishes from both community and government. Dewey's understanding of the public assigns primary importance to indirect consequences of transactions as its constitutive principle.[24] As with any other definition, the problem of grey areas and borderline phenomena exists, and so the consequences must be "lasting, extensive, and serious"[25] if they are to create a public.

An important ramification of this way of thinking about the nature of the public is that material contexts will determine the scope of publics. Dewey explicitly asserts that the extent of a public will vary greatly, depending upon the development of technology: "The consequences of conjoint behavior differ in kind and in range with changes in 'material culture,' especially those involved in exchange of raw materials, finished products, and above all in technology, in tools, weapons, and utensils. These in turn are immediately affected by inventions in means of transit, transportation, and intercommunication."[26] This means that there are potentially many publics and that the size of publics has varied greatly in history.

Dewey provides a radically anticonstructivist and antisociological understanding of the public, and this move sharply poses the problem of a public's relationship to patterns of legitimate authority (which Dewey calls "government") and communal identity (which he calls "community"). Because they are created by indirect consequences, the scope and hence membership of publics are not determined by human intentions, feelings of solidarity, or social traits held in common, but rather by actual, significant, and enduring consequences. Thus, the public is not necessarily equivalent to the members of any actual political community, or subject to the rule of any actual government. The central insight of Dewey's politics is that the discrepancy between the public created by the material forces of the Industrial Revolution and lingering preindustrial forms of government and community underlies the many political problems of the twentieth century.

The temporal and spatial scope of the impacts of modern industrial civilization have created an intergenerational public. Current decisions

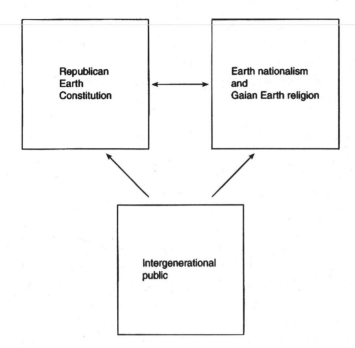

Figure 12.2
Patterns of authority and identity for an intergenerational public

about resource use and abuse — most notably depletion of nonrenewable resources, loss of species and biological diversity, alterations in the Earth's climate, and generation of large quantities of extremely long-lived radioactive and toxic wastes — now promise to affect future generations in ways lasting, extensive, and serious.[27]

Once this transgenerational public is posited as sovereign — the ultimate and undivided source of legitimate authority — the questions arises as to which particular patterns of legitimate authority and communal identity arise from it and could adequately manifest and instantiate it. The simple answer to these questions is: a republican Earth constitution, and Earth nationalism and Gaian Earth religion (see figure 12.2).

A Federal-Republican Earth Constitution

The reason a republican Earth constitution is the form of government consistent with a transgenerational public sovereign is quite straightforward. Because most of the members of the transgenerational public cannot register their preferences in the present, a system of constraints on the living generation of humans is necessary. Because the essence of a republican constitution is a system of legitimate authorities that restrain, such a constitution can best manifest the interests of a trans-generational sovereign public. A republican constitution for the Earth entails a system of restraints that prevents the living from altering the planet in ways that are inconsistent with the fundamental interests of the sovereign, the intergenerational public.

As with all constitutions, the actual stipulations of such a constitution must be a judicious melding of permanent principles with contingent circumstance. Such a constitution requires an intergenerational balance sheet, and an authoritative mechanism to restrain the appetites and actions of the present through a mix of conservation, preservation, and restoration. Such a constitution would itself not be a specific enforceable law, but rather a set of standards against which specific laws and authoritative actions would be held accountable. Such an Earth consti-tution would not require and would not be consistent with a centralized and hierarchical world state or government, because such an entity would amplify the power of one group of the living members of the sovereign intergenerational public not only against the future members of the sovereign entity, but against living ones as well.

Such a constitution would not overthrow or replace the large and growing body of existing environmental domestic laws and regulations and international regimes. Rather, it would constitute the central prin-ciples of more specific authoritative measures, and establish a system for voiding measures and acts inconsistent with its principles.

It is not necessary for such a constitution to be written and established in one temporally contained act of drafting and establishment (or "found-ing"). Given the complexities and uncertainties with which it must cope, it would be impractical and undesirable to attempt such a "top-down" founding. The multiple existing processes of environmental governance formation now under way can be viewed as subcommittee meetings of an Earth Constitutional Convention. A full-fledged constitutional founding is appropriate only when environmental governance thickens and elaborates to the point where questions of overlap, priority, and institutionalization emerge as practical problems best solved by the formulation of general principles crystallized as an architectonic structure of legitimate authority.

Earth Nationality and Earth Religiosity

The expression of an intergenerational public sovereign in a republican Earth constitution provides part of the basis of a terrapolitian order, but poses the fundamental problem of what forms of communal identity could manifest and instantiate this sovereignty. It should be immediately obvious that the problems traditionally plaguing republican constitutional orders and extended popular sovereigns are particularly acute for an intergenerational sovereign public and a republican Earth constitution. The problem of community and group solidarity is severe because the future generations that constitute the majority of the members of the intergenerational public sovereign obviously cannot participate in communal activities with the currently living generation. Similarly, inducing virtue in the sense of self-restraint is particularly difficult in modern industrial societies where production capacities and consumptive appetites have been fully cultivated.

In conceptualizing potential solutions to these problems it is most useful to examine the two powerful social forces that extended sovereigns and republican political orders have traditionally been heavily dependent upon (national identity and religion) and attempt to conceptualize reconfigurations of them (Earth nationality and Earth religion) that are more directly and intimately related to the sustainability imperatives of an intergenerational sovereign public.

Planetary *Topophilia* and Earth Nationality

The specific contents of "national" identities[28] vary greatly, but are commonly built upon one or some combination of three elements: (1) an ethnonational identity as member of a group based upon shared attributes (such as language, history, and religion, etc.) that distinguish members from nonmembers; (2) an identity based upon membership in a particular political community or political regime, which gives rise to a regime patriotism; and (3) an identity and loyalty based upon the experiences and feelings of connectedness to a particular place or area, a sentiment dubbed *geopiety* by John Kirtland Wright[29] and *topophilia* by Yi-Fu Tuan.[30]

The "here-feeling" component of national identity has been much less studied and appreciated by recent students of national identity than the "we-feeling" based on group attributes and differences such as language, religion, history, and institutions.[31] But "here feeling" figures prominently in many of the most seminal political theorists of national identity. In his classic formulation of the sources of patriotism, Edmund Burke speaks of an "instinct" that "extends even to brute creation." This powerful sentiment is "a fondness for the place where they have been bred, for the habitations [they] have dwelt in" that "binds all creatures to their country" and "never becomes inert," and does not "ever suffer us to want a memory of it."[32] Nationalist rhetoric and discourse characteristically claim that national identity is natural or primordial. Assertions about specific places figure prominently in these constructs, as do claims about the character of the relationship between the human and the biological natural, both of "race and blood" and of "soil and land."

The experience of place and sentimental attachments to it can be, and already is being, directed at the Earth as a whole. This constitutes what I have elsewhere analyzed as "Earth nationalism" and "Earth patriotism."[33] This planetary evocation of place is most graphic in the "whole Earth picture," photographs taken of the Earth from outer space. Here the Earth is credibly experienced as a home place that is soft, fluid, and fuzzy. It also seems vulnerable, isolated, and precious. It evokes an aura of a unique and distinctive place.

This "whole Earth" representation is fundamentally different from that of the "globe." A globe is constituted by spherical Cartesian space within which artificial political borders are cartoonishly prominent. In contrast, the whole Earth picture is an actual photograph of what the Earth looks like to those with the most comprehensive positional vantage point. The two most distinctive features of a globe are the presence of a grid of longitude and latitude, and the representation of different nation-states with different colors. Globes are spherical maps, but the whole Earth picture depicts the planet as lacking in sharp lines and angles. And the whole Earth picture does not display the borders of nation-states, because the lines separating state from state are mostly invisible from space. The whole Earth picture thus seems more authentic and "real" than the globe, whose spatial representations seem constructed, arbitrary, unnatural, and conventional—fictions and conceits imposed upon reality or mistaken for reality rather than what is real.[34]

A crucial ingredient in the experience of planetary *topophilia* not present in earlier forms of "place-feeling" is the role of modern natural science. Early European nationalist intellectuals and ideologues such as Rousseau, Burke, and Herder attacked the universalistic and rationalistic science of the Enlightenment and championed folk traditions and identities shaped by primitive material circumstances. In contrast, contemporary planetary environmental awareness contains a major element of ecological and Earth systems natural science. Where previous nationalism employs a preecological or antiecological understanding of place and environment, the emergent Earth nationalism integrates scientific ecological understandings of place and human links to place.

A second distinctive and politically significant feature of contemporary environmentalist evocations of place sensibility is that they are diverse, layered, and overlapping. The ideologies of the modern nation-state assert "one place, one people," and proceed to homogenize diverse places and diverse peoples. The "nation building" of modern states usually entails extinguishing or marginalizing the group identities of diverse peoples unfortunate enough to be caught within the internationally recognized borders of a state apparatus with "modern" ambitions.[35]

In contrast, topophilic environmentalists assert the existence of place claims that are diverse and overlapping as well as distinct. Bioregionalist theory and practice asserts that political borders and identities should reflect distinct natural bioregions. But the borders of different watersheds and ecosystems do not form sharp lines, but rather overlay one another in a complex pattern. Furthermore, the unmistakable political message of ecological science is that the entirety of the Earth is the only fully integral bioregion, and that the "homeland" of all humans is the planet and distinct but overlapping pieces of it.[36] Thus, the ideologies of Earth nationality in essence assert "one diverse place, one diverse people." This layered *topophilia* of contemporary bioregionalist theory and practice cannot be politically instantiated by an altered configuration of bionational states, but rather requires federal arrangements.

Sacred Earth and Gaian Religiosity
The last and possibly most important component of a communal identity congruent with the needs of a transgenerational public sovereign

is religion, a topic that is arguably more complex and problematic than all the other issues addressed thus far.[37] As noted, contemporary environmental practice and theory have involved riotous conceptual and practical ferment about religion and its relationship to Earth sustainability. Despite the disdain of secular intellectuals and academics, Earth spirituality is ubiquitous in the popular environmental movement. A good index of how strongly this current runs is the prominent role spiritual concerns play in Al Gore's widely read *Earth in the Balance*.[38] Much of this effort has focused upon the relationship between the existing major religious traditions and Earth sustainability.[39] Earth religiosity also plays a prominent role in the radical environmental movement.[40] A substantial effort has also been made to resurrect older forms of Earth spirituality and religion and to construct Earth religious cosmologies and theologies, as well as ceremonial and ritual practices, consistent with modern Earth system science. A particularly rich source for these efforts are the "living fossils" of marginalized "first nations" and "indigenous peoples."[41] Among the more radical and constructive of these Earth theologies is what might be termed *Gaian Earth religion*, which is also the most potentially useful for instantiating communal identities congruent with the sustainability imperatives of an intergenerational public sovereign.

In approaching this issue, our central concern is not the substantive merits of the theological, cosmological, or scientific claims of Gaian Earth religion, but rather its potential role as a source of communal identity consistent with an intergenerational public sovereign. Analysis of the functional fit between religions and political orders has a distinguished lineage. Many political theorists have astutely analyzed the relationship between specific religions and specific political orders. For example, Machiavelli famously argued that Roman paganism was a crucial feature in the Roman republic.[42] More recently, John Stuart Mill argued for the utility of theism as a ground for morality.[43] Others, most notably Hobbes and Rousseau, have gone further and sketched the features of hypothetical religions that they argue are best suited to establishing and maintaining their visions of a necessary or preferred political order.[44] These analyses and constructions have one thing in common: They are not based upon a love or reverence for God, but rather seek to use religion as an instrument for creating and maintaining particular political orders.[45]

There is nothing about Earth religiosity in general, and Gaian Earth religion in particular, that is not subject to fundamental flux and contestation. But two general features of the emergent Earth religiosity stand out as interesting for their potential roles in a terrapolitan "civic religion" and sources of Earth sustainable communal identities.

First, Earth religion is a relative rarity—a modern worldview with a scientifically credible cosmology.[46] A major limitation of the great premodern theological cosmologies is that modern natural science has undermined their credibility. In the West the recession of Christendom and the rise of modernity were marked by scientific discoveries profoundly subversive of the authoritative religious cosmology: the Copernican displacement of the Earth from the center of the universe,[47] the discovery of deep geological time, the theory of biological evolution, and the emergence of a mechanical conception of nature hostile to supernatural intervention. From the deistic recasting of God into a retired watchmaker it was but a short step to Nietzsche's "death of God" and the decline of monotheistic religion as the axis of identity and community.

A striking feature of Gaian Earth religion as a spiritual and moral system is its ability to make at least a prima facie claim to being compatible with the important natural science of ecology.[48] Gaia is the most salient metaphorical structure spanning the divide between ecological science and Earth identity narratives. *Gaia* is the term employed by the Earth system scientist James Lovelock for the comprehensive homeostatic system of the planet's interacting living organisms and geophysical features.[49] For theologians *Gaia*, denotes an encompassing spiritual reality grounded in nature.[50] Lovelock, after observing that "a separation of life into sacred and secular parts" is not plausible, articulates the central claim of the Earth science-religion fusion: "Thinking of the Earth as alive makes it seem, on happy days, in the right places, as if the whole planet were celebrating a sacred ceremony. Being on the Earth brings that same special feeling of comfort that attaches to the celebration of any religion when it is seemly and when one is fit to receive."[51]

Lovelock's carefully hedged formulation acknowledges the existence of a very old and widely encountered "variety of religious experience." The derivation of norms and prescriptions from bodies of modern scientific knowledge remains a deeply problematic undertaking, but when linked with a simple set of normative assumptions, particularly the desirability of survival, ecology comes much closer to seeming to provide a set of broad and important norms.

A second central feature of the new and radical forms of Earth religiosity is the assertion of special sacred places, emotively powerful ceremonies and rituals,[52] and cosmological narratives drawn with primordial metaphorical resources and potentially open to wide vernacular accessibility.[53] Premodern religious traditions generated hierarchical space in which special locations were deemed particularly sacred and hallowed. These sacred places served as gathering places for ceremonial rituals and for spiritual encounters with the extrahuman forces giving direction and meaning to human existence. A strong version of this sense of sacred space is found in the wilderness "sanctuary."[54] In this "church of the Earth" the great "wonders of nature" constitute the cathedrals and sacred grounds of nature. These sites evoke powerful emotive experiences, and are the destinations of mass pilgrimages, similar to those of Islam and Roman Catholicism.

Judged by functional criteria, Gaian Earth religiosity seems well suited to serve as the "civic religion" for a federal-republican Earth constitution. It potentially could underpin the social norms and behaviors of restraint that are necessary to achieve a sustainable society, but which are very difficult to support on their own right. The effects of many environmental problems are most likely to be felt in the future, or in distant places, while the tasks necessary to achieve a sustainable society involve real, immediate sacrifice and must be performed routinely by vast numbers of people. Reason and appeals to higher self-interest or long-run self-interest may be insufficient to motivate sufficient action.[53] The appeal of Earth religion is that it helps motivate behavior respectful of the Earth which otherwise would be difficult to achieve, by providing a system of meaning that can span generations and foster a sense of transgenerational communal identity.

Conclusions

This brief sketch of the logic of the structures of authority and communal identities consistent with the assertion of the sovereignty of an intergenerational public raises many fundamental conceptual issues that require further investigation. But to close this first excursus, it is useful to connect the argument advanced here with the practical strategies of the worldwide environmental movement, and suggest several ways in which the assertion of "green sovereignty" can sharpen and strengthen environmentally responsible political practice.

First, the assertion of a green sovereign and its correlative governance and identity programs can add a degree of coherence and unity to the worldwide environmental movement, which is marked by extreme diversity and fragmentation. The assertion of popular sovereignty added coherence and legitimacy to the political emergence of the masses in modern Europe and elsewhere. In a similar manner, the assertion of green sovereignty can help raise the environmentalist agenda to the fundamental political and cultural challenge that it in fact is.

Second, the assertion of green sovereignty can help strengthen and legitimate the practices and discourses of virtue that comprise a powerful part of contemporary environmental politics. Throughout the advanced industrial world, and particularly in the United States, neoconservatives have vigorously reasserted the importance of various forms of virtue in liberal society and have forced a wide-ranging reexamination of the ways in which government policies undermine or foster individual self-restraint.[56] Notably absent in the tablets of value advanced by partisans of neoconservative "virtuepolitik" has been consumptive and reproductive self-restraint. The "voluntary simplicity" practice and discourse of contemporary environmentalists constitutes a powerful, but underpoliticized challenge to the deadly hidden permissiveness of the neoconservative values agenda.[57] Given that capitalism has now achieved something approaching an "end of history" ideological hegemony, sovereignty rests not with states, but with the community of consumers—the mythical "consumer sovereignty" of neoclassical economics. The green virtues of voluntary simplicity constitute a radical assault on the sovereignty of consumptive appetites.

A hidden permissiveness also characterizes the dominant culture's reproductive norms. The major organized resistance to reproductive responsibility is squarely grounded in religious claims. Despite the clear links between burgeoning human populations, amplified by technology and affluence, and environmental degradation, the reproductive permissiveness of the various religious "spiritual humanists" has not been challenged at a fundamental level. The reproductive rate of the "'virtuous" Mormons rivals that of the Third World, and the Roman Catholic hierarchy continues to proscribe all forms of artificial contraception and family planning. A powerful international coalition of Islamic traditionalists, Roman Catholics, Evangelical Protestants, and African patriarchal animists has emerged to block the further expansion, funding, and legitimacy of international family planning programs. By grounding reproductive responsibility in the assertion of an Earth religion, opponents of reproductive permissiveness will gain a powerful new political tool.

Third, the assertion of the sovereignty of an intergenerational public can add a potentially powerful component to the environmental challenge to the sanctity of state sovereignty in world politics. Instead of challenging sovereignty as the basis of authority, the assertion of a green sovereignty can put the claims of environmentally sound practices on a more secure footing.

Notes

1. An earlier version of this paper was presented at the ISA-West, October 1996. Richard Matthew, Bron Taylor, and Paul Wapner offered helpful comments on earlier versions.

2. Hedley Bull, "The State's Positive Role in World Affairs," *Daedalus* 108, no. 4 (Fall 1979): 101–10; and Barry Buzan, "The Timeless Wisdom of Realism?," in Steve Smith, Ken Booth, and Marysia Zalewski, eds., *International Theory: Positivism and Beyond* (Cambridge: Cambridge University Press, 1996): 47–65.

3. The most sustained and systematic indictment has been made by members of the World Order Models Project. See Richard Falk, *On Humane Governance: Toward a New Global Politics* (University Park: Pennsylvania State University Press, 1995): 79–103; and Joseph A. Camilleri and Jim Falk, *The End of Sovereignty?* (Aldershot, England: Elgar, 1992).

4. James Rosenau, "The State in an Era of Cascading Politics: Wavering Concept, Widening Competence, Withering Colossus, or Weathering Change?," in James Caporaso, ed., *The Elusive State* (London: Sage, 1989): 17–49; Janice Thomson and S. Krasner, "Global Transactions and the Consolidation of Sovereignty," in Ernst-Otto Czempiel and James Rosenau, eds., *Global Changes and Theoretical Challenges* (Lexington, Mass.: Lexington Books, 1989); and Stephen D. Krasner, "Compromising Westphalia," *International Security* 20, no. 3 (Winter 1995–96): 115–51.

5. The most powerful treatment of the central role of limits and constraints in

sustainable governance is found in Garrett Hardin, *Living within Limits: Ecology, Economics, and Population Taboos* (New York: Oxford University Press, 1993).

6. Karen Litfin, "Ecoregimes: Playing Tug of War with the Nation-State System," in Ronnie Lipschutz and Ken Conca, eds., *The State and Social Power in Global Environmental Politics* (New York: Columbia University Press, 1993): 94–117.

7. Oran Young, *International Cooperation: Building Regimes for Natural Resources and the Environment* (Ithaca, N.Y.: Cornell University Press, 1989); and Abram Chayes and Antonia Handler Chayes, *The New Sovereignty: Compliance with International Regulatory Agreements* (Cambridge, Mass.: Harvard University Press, 1996).

8. For example, Jean Bethke Elshtain calls for a "postsovereign politics" in "Sovereignty, Identity, Sacrifice," *Social Research* 58, no. 3 (Fall 1991): 560. See also James Rosenau's celebration of the growing importance of "sovereignty-free" actors in world politics, in Rosenau, *Turbulence in World Politics* (Princeton, N.J.: Princeton University Press, 1990).

9. David Macauley, ed., *Minding Nature: The Philosophers of Ecology* (New York: Guilford: 1996); Martin O'Connor, ed., *Is Capitalism Sustainable? Political Economy and the Politics of Ecology* (New York: Guilford Press, 1994); Adrian Atkinson, *Principles of Political Ecology* (London: Belhaven Press, 1991); Fritz Capra, *Green Politics* (New York: Simon and Schuster, 1984); Andrew Dobson, *Green Political Thought*, 2d ed. (London: Routledge, 1990, 1995); John Dryzek, *Rational Ecology: Environment and Political Economy* (Oxford: Basil Blackwell, 1987); Robert E. Goodin, *Green Political Theory* (London: Polity, 1992); Adolf G. Gunderson, *The Environmental Promise of Democratic Deliberation* (Madison: University of Wisconsin Press, 1995); William Ophuls, *Ecology and the Politics of Scarcity: Prologue to a Political Theory of the Steady State* (San Francisco: W. H. Freeman, 1977); Bob Pepperman Taylor, *Our Limits Transgressed: Environmental Political Thought in America* (Lawrence: University of Kansas Press, 1992).

10. F. H. Hinsley, *Sovereignty*, 2d ed. (Cambridge: Cambridge University Press, 1986).

11. Roderick Nash, *The Rights of Nature* (Madison: University of Wisconsin Press, 1989).

12. Matthew Fox, *The Coming of the Cosmic Christ: The Healing of Mother Earth and the Birth of a Global Renaissance* (San Francisco: HarperCollins, 1988); Eugene C. Hargrove, ed., *Religion and the Environmental Crisis* (Athens: University of Georgia Press, 1986); J. Baird Callicott, *Earth's Insights: A Multicultural Survey of Ecological Ethics from the Mediterranean Basin to the Australian Outback* (Berkeley: University of California Press, 1994); and Theodore Roszak, *The Voice of the Earth* (New York: Simon and Schuster, 1992).

13. William Blackstone, *Commentaries on the Laws of England* (Oxford: Clarendon Press, 1765–69), vol. 1, 156–57.

14. Edmund S. Morgan, *Inventing the People: The Rise of Popular Sovereignty in England and America* (New York: W. W. Norton, 1988).

15. James Otis, "The Rights of the British Colonies Asserted and Proved" (Boston, 1764).

16. For extended analysis of this point, see Daniel Deudney, "The Philadelphian System: Sovereignty, Arms Control, and Balance of Power in the American States-Union, ca. 1787–1861," *International Organization* 49, no. 2 (Spring 1995): 191–228.

17. For the relationship between constraints on governmental power and constitutionalism, see Frederick Hayek, *The Constitution of Liberty* (Chicago: University of Chicago Press, 1960); and Jon Elster and Rune Slagstad, eds., *Constitutionalism and Democracy* (Cambridge: Cambridge University Press, 1988).

18. Stephen Holmes, *Passions and Constraint: On the Theory of Liberal Democracy* (Chicago: University of Chicago Press, 1995).

19. The strongly gendered formulation of republican communal identity (*virtue* from *vir* — Latin for male human being) is a product of the fact that popular citizenship was historically first possible only in small — and therefore militarily precarious — polities that were compelled by these circumstances to cultivate martial virtue (what Machiavelli called *virtu*) in an armed male citizen class.

20. Wilson Carey McWilliams, *The Idea of Fraternity in America* (Berkeley: University of California Press, 1973).

21. War patriotism played a particularly important role in reinforcing liberal communal identity in the United States, and was able to do so because of the historical accident that the major military adversaries of the United States have always been decidedly less liberal than the United States.

22. Among the voluminous literature, two small pieces provide useful overviews: James D. Savage, "Corruption and Virtue at the Constitutional Convention," *Journal of Politics* 56, no.1 (February 1994): 174–86; and J. Peter Euben, "Corruption," in Terence Ball, James Farr, and Russell Hanson, eds., *Political Innovation and Conceptual Change* (Cambridge: Cambridge University Press, 1989): 220–46.

23. John Dewey, *The Public and Its Problems* [1927] (Chicago: Swallow Press, 1954): 15–16.

24. The opposite of the public is the private: "When the consequences of an action are confined, or are thought to be confined, mainly to the persons directly engaged in it, the transaction is a private one." Dewey, *Public and Its Problems*, 18–19. This formulation is free of the assumptions about gender and domesticity that frequently mar analyses of the differences between public and private.

25. Dewey, *Public and Its Problems,* 67.

26. Dewey, *Public and Its Problems,* 44.

27. Edith Brown Weiss, *In Fairness to Future Generations: International Law, Common Patrimony, and Intergenerational Equity* (New York: Transnational Publishers and United Nations University, 1988); and John Gerard Ruggie, "International Structure and International Transformation: Space, Time, and Method," in Czempiel and Rosenau, *Global Changes and Theoretical Challenges: Approaches to World Politics for the 1990s* (Lexington, Mass.: Lexington Books, 1989).

28. For the diverse ingredients of "national" identities, see Anthony D. Smith, *Theories of Nationalism* (New York: Harper and Row, 1971); and Benedict Anderson, *Imaged Communities: Reflections on the Origins and Spread of Nationalism,* 2d ed. (London: Verso, 1991).

29. Yi-Fu Tuan, "Geopiety: A Theme in Man's Attachment to Nature and to Place," in David Lowenthal and Martyn J. Bowden, eds., *Geographies of the Mind* (New York: Oxford University Press, 1976).

30. Yi-Fu Tuan, *Topophilia: A Study of Environmental Perception, Attitudes, and Values* (New York: Columbia University Press, 1974).

31. Anthony D. Smith, "Poetic Spaces: Uses of Landscape," in *The Ethnic Origins of Nations* (Oxford: Basil Blackwell, 1986): 183–91.

32. Edmund Burke, *The Writings and Speeches of the Right Honourable Edmund Burke* (Boston: Little, Brown, 1901): 11:422–23.

33. Daniel Deudney, "Ground Identity: Nature, Place, and Space in Earth Nationalism," in Yosef Lapid and Friedrich Kratochwil, eds., *The Return of Culture and Identity in IR Theory* (Boulder, Colo.: Reinner, 1995): 129–45.

34. For contestations over the meaning and appropriations of "whole Earth" imagery, see Guy Geney, "Gaia: The Globalitarian Temptation," in Wolfgang Sachs, ed., *Global Ecology* (London: Zed, 1993); Denis Cosgrove, "Contested Global Visions: One-World, Whole-Earth, and the Apollo Space Photographs," *Annals of the Association of American Geographers* 84, no. 2 (1994): 270–94; Yaakov Jerome Garb, "The Use and Misuse of the Whole Earth Image," *Whole Earth Review* (March 1995); and Wolfgang Sachs, "The Blue Planet: An Ambiguous Modern Icon," *The Ecologist* 24, no. 5 (September-October 1995).

35. Jason Clay, "Resource Wars: Nation and State Conflicts in the Twentieth Century," in Barbara Rose Johnston, ed., *Who Pays the Price?* (Washington, D.C.: Island Press, 1994):19–30; and Bernard Nietschmann, "The Fourth World: Nations vs. States," in George J. Demko and William B. Wood, eds., *Reordering the World: Geopolitical Perspectives on the Twenty-First Century* (Boulder, Colo.: Westview, 1994): 225–42.

36. Sam Love, "Redividing North America," *The Ecologist* 7, no. 7 (September 1977): 318–19; Ernest Callenbach, *Ectopia: The Notebooks and Reports of William Weston* (Berkeley: Banyon Tree, 1975); Ernest Callenbach, *Ectopia Emerging* (Berkeley: Banyon Books, 1981); Kirkpatrick Sale, *Dwellers in the*

Land: The Bioregional Vision (San Francisco: Sierra Club Books, 1985); and Van Andrews, et al., eds., *Home! A Bioregional Reader* (Philadelphia: New Society Publishers, 1990).

37. For the classic statement of the religious-political question and the distinctively modern Western answer to it, see Leo Strauss, *Spinoza's Critique of Religion* (New York: Schocken Books, 1965).

38. Al Gore, *Earth in the Balance: Ecology and the Human Spirit* (New York: Houghton Mifflin, 1992).

39. For overviews, see Robert Booth Fowler, *The Greening of Protestant Thought* (Chapel Hill: University of North Carolina Press, 1995).

40. Bron Taylor, "The Religion and Politics of Earth First!," *The Ecologist* 21, no. 6 (November 1992): 258–66.

41. For the controversies surrounding these appropriations, see Bron Taylor, "Earthen Spirituality or Cultural Genocide?: Radical Environmentalism's Appropriation of Native American Spirituality," *Religion* 27 (Spring 1997): 183–215.

42. Niccolo Machiavelli, *The Discourses*, Walker and Richardson, trans. (Harmondsworth: Penguin, 1970), book 1, sections, 11–15, 139–52.

43. John Stuart Mill, *Nature: The Utility of Religion and Theism*, 2d ed. (London: Longmans, Green, Reader, and Dyer, 1824).

44. Terrence Ball, "Rousseau's Civil Religion Reconsidered," and "The Survivor and the Savant: Two Schemes for Civil Religion Compared," *Reappraising Political Theory* (Oxford: Clarendon Press, 1995).

45. "Religion is being recommended to us because it supports morality, not morality because it derives from religion." Joan Robinson, *Economic Philosophy* [1962] (Harmondsworth, England: Penguin, 1964): 15.

46. An earlier effort in this direction by the Roman Catholic theologian Teilhard de Chardin was based on views of physics that have been subsequently superseded.

47. Alexander Koyre, *From Closed World to Infinite Universe* (Baltimore: Johns Hopkins University Press, 1957).

48. David Oates, *Earth Rising: Ecological Belief in an Age of Science* (Corvallis: Oregon State University Press, 1984); and William Irwin Thompson, *Gaia: A Way of Knowing: Political Implications of the New Biology* (Great Barrington, Mass.: Intertraditions/Lindisfarne, 1987).

49. James Lovelock, *Gaia: A New Look at Life on Earth* (New York: Oxford University Press, 1979); James Lovelock, *The Ages of Gaia: A Biography of Our Living Earth* (New York: Norton, 1988); and James Lovelock, *Healing Gaia: Practical Medicine for the Planet* (New York: Harmony, 1991).

50. Dorian Sagan and Lynn Margulis, "Gaian Views," in Christopher Key Chapple, ed., *Ecological Prospects: Scientific, Religious, and Aesthetic Perspectives* (Albany: State University of New York Press, 1994); Rosemary Radford Ruether, *Gaia and God: An Ecofeminist Theology of Earth*

Healing (New York: HarperCollins, 1992); and Anthony Weston, "Forms of Gaian Ethics," *Environmental Ethics* 9 (Fall 1987): 217–30.

51. Lovelock, *The Ages of Gaia*, 204–5.

52. Dolores LaChapelle, *Earth Wisdom* (Silverton, Colo.: Kivaki Press 1978); and Dolores LaChapelle, *Earth Festivals* (Silverton, Colo.: 1976).

53. Max Oelschlaeger, "The Sacred Canopy: Religion as Legitimating Narrative," chapter 3, and "Redescribing Religious Narrative: The Significance of the Sacred Story," chapter 6, in *Caring for Creation: An Ecumenical Approach to the Environmental Crisis* (New Haven: Yale University Press, 1994).

54. Linda H. Graber, *Wilderness as Sacred Space* (Washington, D.C.: Association of American Geographers, 1976).

55. For sustained analysis of this point, see Fred Hirsch, "The Moral Reentry," chapter 10, *Social Limits to Growth* (Cambridge, Mass.: Harvard University Press, 1976): 137–51.

56. William J. Bennett, ed., *The Book of Virtues: A Treasury of Great Moral Stories* (New York: Simon and Schuster, 1995); and Mary Ann Glendon and David Blankenhorn, eds., *The Seedbeds of Virtue: Sources of Competence, Character, and Citizenship in American Society* (Lanham, MD: Madison Books, 1995).

57. Duane Elgin, *Voluntary Simplicity* (New York: William Morrow, 1981).

Index